WELCOME TO THE SPACE.com COLLECTION

Get ready to explore the wonders of our incredible universe. The Space.com Collection is packed with amazing astronomy, incredible discoveries and the latest missions from space agencies around the world. From distant galaxies, to the planets, moons and asteroids of our own solar system – you'll discover a wealth of facts about the cosmos, and learn about the new technologies, telescopes and rockets in development that will reveal even more of its secrets. Space.com launched 20 years ago and fast became the premier source of space exploration, innovation and astronomy news, chronicling – and celebrating – humanity's ongoing expansion across the final frontier. For us, exploring space is as much about the journey as it is the destination.

FUTURE

SPACE.com COLLECTION

Future PLC Quay House, The Ambury, Bath, BA1 1UA

Bookazine Editorial
Editor **Jacqueline Snowden**
Designer **Steve Dacombe**
Compiled by **Katharine Marsh & Katy Stokes**
Editorial Director **Jon White**
Senior Art Editor **Andy Downes**

Space.com Editorial
VP of Content and Global Editor-in-Chief **Bill Gannon**
Managing Editor **Tariq Malik**

Cover images
NASA, ESA, Thinkstock

Photography
NASA, ESA, Thinkstock, Getty Images
All copyrights and trademarks are recognised and respected

Advertising
Media packs are available on request
Commercial Director **Clare Dove**

International
Head of Print Licensing **Rachel Shaw**
licensing@futurenet.com
www.futurecontenthub.com

Circulation
Head of Newstrade **Tim Mathers**

Production
Head of Production **Mark Constance**
Production Project Manager **Matthew Eglinton**
Advertising Production Manager **Joanne Crosby**
Digital Editions Controller **Jason Hudson**
Production Managers **Keely Miller, Nola Cokely, Vivienne Calvert, Fran Twentyman**

Printed by William Gibbons, 26 Planetary Road, Willenhall, West Midlands, WV13 3XT

Distributed by Marketforce, 5 Churchill Place, Canary Wharf, London, E14 5HU
www.marketforce.co.uk Tel: 0203 787 9001

Space.com Collection Third Edition (ASB3731)

Future plc is a public company quoted on the London Stock Exchange (symbol: FUTR)
www.futureplc.com

Chief executive **Zillah Byng-Thorne**
Non-executive chairman **Richard Huntingford**
Chief financial officer **Rachel Addison**

Tel +44 (0)1225 442 244

Part of the SPACE.com bookazine series

CONTENTS

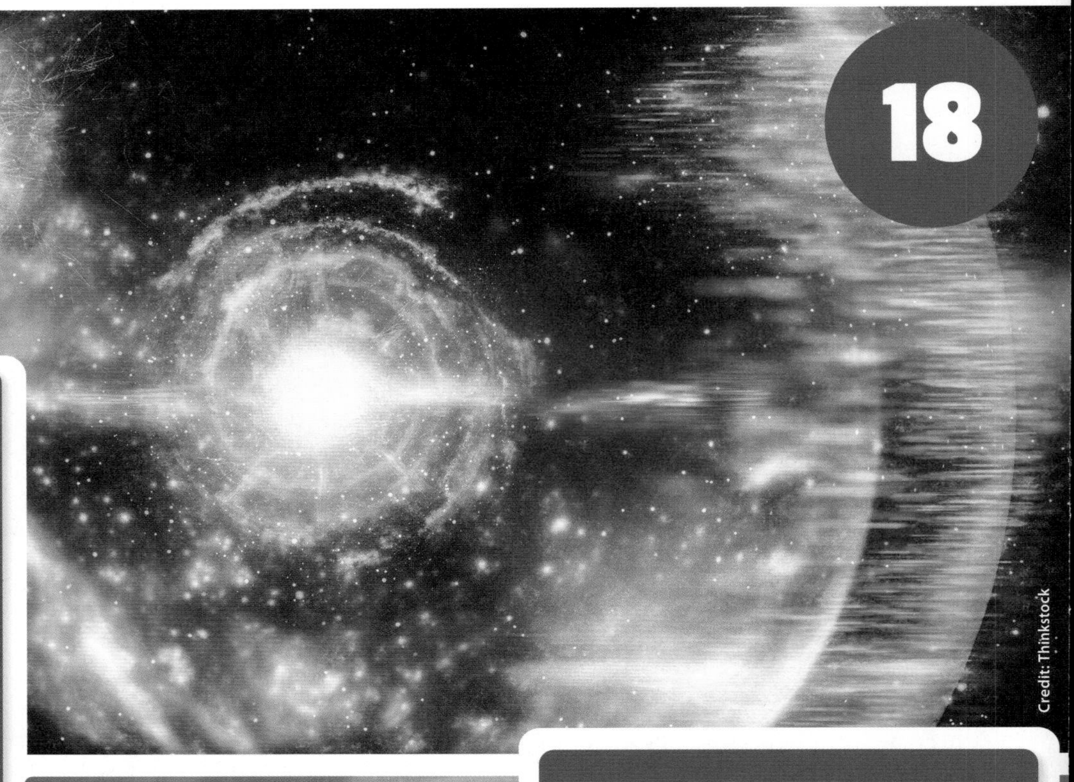

Credit: Thinkstock

THE UNIVERSE AND ITS ORIGINS

Credit: Thinkstock

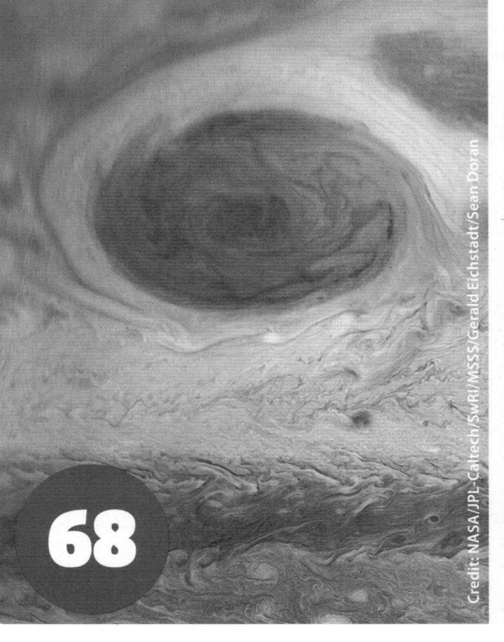

Credit: NASA/JPL-Caltech/SwRI/MSSS/Gerald Eichstadt/Sean Doran

THE SOLAR SYSTEM

CONTENTS

Credit: ESA/Hubble, NASA, M. Kornmesser

COSMIC PHENOMENA

102

Credit: NASA

EXPLORATION

Credit: NASA/JPL-Caltech/R. Hurt (SSC-Caltech)

MOST AMAZING SPACE PHOTOS OF 2021

HERE IS SPACE.COM'S SELECTION OF SOME OF THE BEST ASTROPHOTOGRAPHY OF THE YEAR SO FAR...

Credit: ESA/Hubble & NASA, ESO, J. Lee and the PHANGS-HST Team

In this breathtaking image captured by the Hubble Space Telescope, the bright heart of the "starburst" galaxy M61 shines, surrounded by outstretched, winding arms. M61 is classified as a starburst galaxy because of its bright, glowing spots of star formation that can be seen as "rubies" of reddish light in this image.

Hubble spotted this lenticular galaxy, a cross between a spiral and elliptical-shaped galaxy, known as NGC 1947. The galaxy, which was originally discovered over 200 years ago, can only be viewed from the southern hemisphere and can be found in the Dorado (Dolphinfish) constellation about 40 million light-years away from Earth.

Credit: ESA/Hubble & NASA, D. Rosario; Acknowledgment: L. Shatz

AMAZING SPACE PHOTOS

In this image, taken by the Color and Stereo Surface Imaging System (CaSSIS) onboard the European Space Agency's ExoMars Trace Gas Orbiter, shows the southeast wall of a small crater on Mars. The crater is found just about a couple hundred miles away from Hellas, a giant impact crater on the planet's surface. This smaller crater stretches about 7.5 miles (12 kilometers) across. The orbiter's photo shows a wide range of colors, which indicate the presence of different minerals in the planet's surface material.

The European Space Agency's Copernicus Sentinel-2 mission snapped this chilly photo of New York City on Feb. 4, 2021 showing the city blanketed in snow. This recent snow storm was classified as "major" and affected a majority of the Northeast United States, with New York declaring a state of emergency for both the immense snowfall and blistering winds.

Copernicus Sentinel-2 is an Earth-observing mission made up of two satellites: Sentinel-2A and Sentinel-2B. The pair monitor and image our planet, orbiting it from space.

In this out-of-this-world selfie, NASA astronaut Mike Hopkins held his camera out and snapped a photo of himself during a spacewalk with fellow NASA astronaut Victor Glover on Feb. 1. "Ever wonder what an astronaut sees when out on a spacewalk? This selfie shows my view reflecting off of my visor. Takes your breath away!" Hopkins wrote on Twitter, where he shared the space selfie.

Credit: Mike Hopkins/NASA

Credit: International Space Station/Twitter

ABOVE: These images, taken from the International Space Station, show Earth's glowing, colorful aurora alongside lights coming from the cities on our planet's surface down below. Aurora is a natural phenomenon in which colorful lights in the sky, which often appear as green, red, yellow or white, are displayed when electrically-charged particles from the sun interact with gases like oxygen or nitrogen in our planet's atmosphere.

Credit: ArianeGroup/ Frank T. Koch / Hill Media GmbH

The first complete upper stage of the Ariane 6 launch vehicle from the European Space Agency is seen here packed into a container to travel from ArianeGroup in Bremen in Germany to the DLR German Aerospace Center in Lampoldshausen, Germany. There, it will undergo hot fire testing, or tests during which all engines are ignited while the launch vehicle remains stationary. These tests, which will take place in near-vacuum conditions, will help to prove that the vehicle is flight ready.

BELOW: This stunning, sandy, sienna-hued landscape is the Tanezrouft Basin (a desolate region of the Sahara Desert) as seen by the Copernicus Sentinel-2 from space. The extremely arid plain is home to scorching temperatures, little water and vegetation and has even been nicknamed the "Land of Terror." This image was captured as part of Copernicus Sentinel-2, a two-satellite mission that is part of the European Space Agency's Copernicus program.

The galaxy NGC 6946, nicknamed "the Fireworks Galaxy," can be seen in this stunning image from the NASA/ESA Hubble Space Telescope. The galaxy got its explosive nickname because, while our Milky Way galaxy has an average of just 1-2 supernovas per century, NGC 6946 has had 10 in the last century.

"The Fireworks Galaxy," the structure of which is somewhere between a full spiral and a barred spiral, can be found 25.2 million light-years from Earth on the border of the constellations Cepheus and Cygnus.

Credit: X-ray: NASA/CXO/University College London/W. Dunn et al; Optical: W.M. Keck Observatory

Do you see the butterfly? This dazzling image of what looks like a red member of the lepidoptera order is actually a nebula in space about 1,400 light-years from our sun. The nebula, officially called Westerhout 40 (W40) is a vast cloud of gas where baby stars can be born. NASA's Spitzer Space Telescope captured this view with its Infrared Array Camera, using three different wavelengths that lend the image its distinct colors. Stars show up in brilliant blue light, while organic molecules are visible as reddish hues. Dusty material around stars show up as yellow and red.

THE UNIVERSE AND ITS ORIGINS

Credit: NASA Goddard Space Flight Center

Credit: ESA

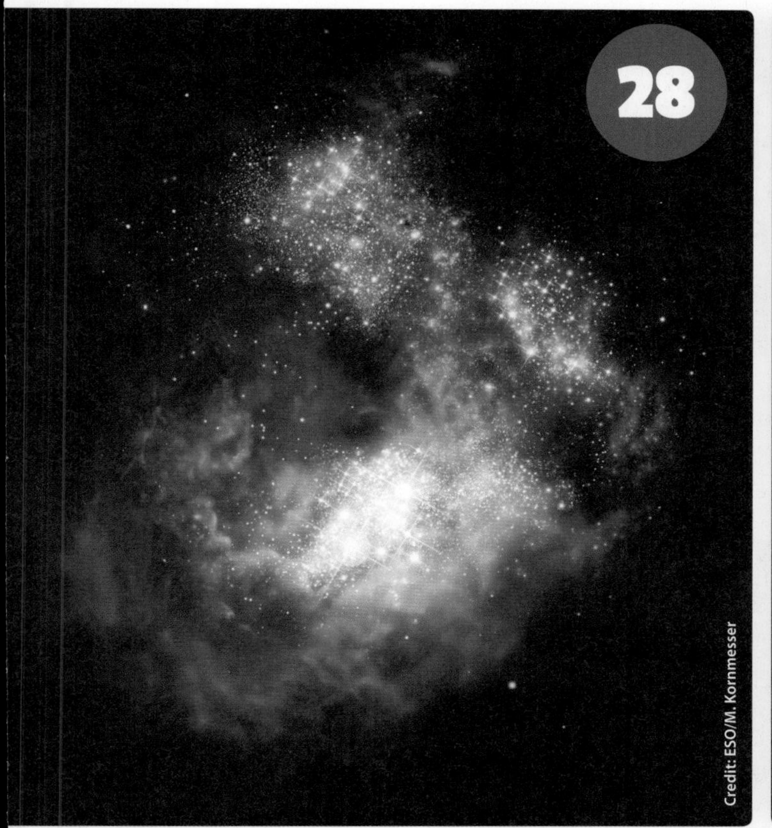
28
Credit: ESO/M. Kornmesser

30
Credit: ESO/M. Kornmesser

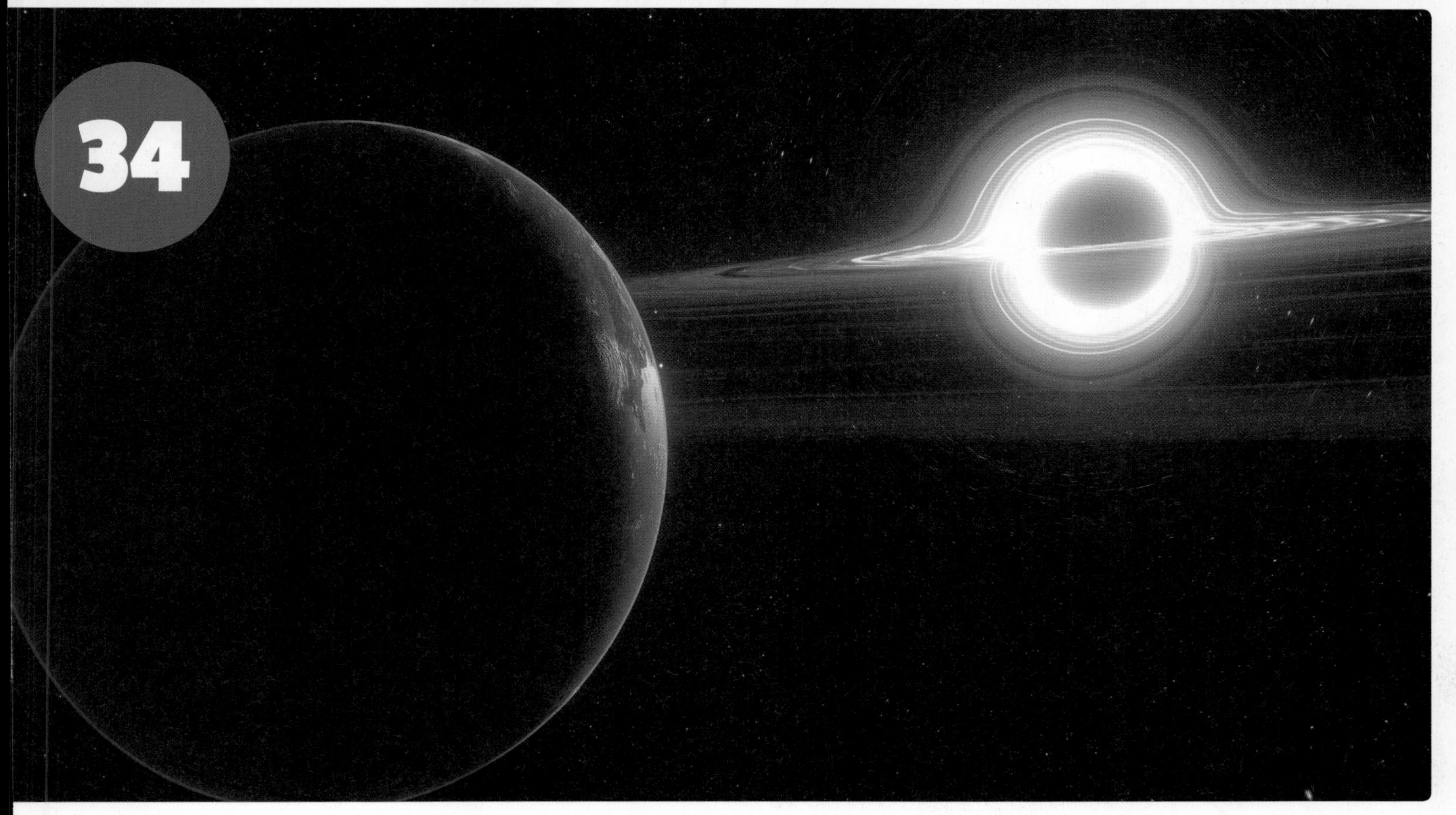
34

WHAT IS THE BIG BANG THEORY?

HOW SCIENTISTS EXPLAIN THE BEGINNINGS OF OUR UNIVERSE

WORDS: ELIZABETH HOWELL

The Big Bang theory is the leading explanation about how the universe began. At its simplest, it says the universe as we know it started with a small singularity, then inflated over the next 13.8 billion years to the cosmos that we know today.

Because current instruments don't allow astronomers to peer back at the universe's birth, much of what we understand about the Big Bang theory comes from mathematical formulas and models. Astronomers can, however, see the "echo" of the expansion through a phenomenon known as the cosmic microwave background.

While the majority of the astronomical community accepts the theory, there are some theorists who have alternative explanations besides the Big Bang – such as eternal inflation or an oscillating universe.

The phrase "Big Bang theory" has been popular among astrophysicists for decades, but it hit the mainstream in 2007 when a comedy show with the same name premiered on CBS. The show follows the home and academic life of several scientists – including an astrophysicist.

A visual representation of the big bang and the expansion of the universe

"ASTRONOMERS CAN SEE THE 'ECHO' OF THE EXPANSION THROUGH A PHENOMENON KNOWN AS THE COSMIC MICROWAVE BACKGROUND"

THE FIRST SECOND, AND THE BIRTH OF LIGHT

In the first second after the universe began, the surrounding temperature was about 10 billion degrees Fahrenheit (5.5 billion Celsius), according to NASA. The cosmos contained a vast array of fundamental particles such as neutrons, electrons and protons. These decayed or combined as the universe got cooler.

This early soup would have been impossible to look at, because light could not carry inside of it. "The free electrons would have caused light (photons) to scatter the way sunlight scatters from the water droplets in clouds," NASA stated. Over time, however, the free electrons met up with nuclei and created neutral atoms. This allowed light to shine through about 380,000 years after the Big Bang. This early light – sometimes called the "afterglow" of the Big Bang – is more properly known as the cosmic microwave background (CMB). It was first predicted by Ralph Alpher and other scientists in 1948, but was found only by accident almost 20 years later.

Arno Penzias and Robert Wilson, both of Bell Telephone Laboratories in Murray Hill, New Jersey, were building a radio receiver in 1965 and picking up higher-than-expected temperatures, according to NASA. At first, they thought the anomaly was due to pigeons and their dung, but even after cleaning up the mess and killing pigeons that tried to roost inside the antenna, the anomaly persisted.

Simultaneously, a Princeton University team (led by Robert Dicke) was trying to find evidence of the CMB, and realised that Penzias and Wilson had stumbled upon it. The teams each published papers in the Astrophysical Journal in 1965.

DETERMINING THE AGE OF THE UNIVERSE

The cosmic microwave background has been observed on many missions. One of the most famous space-faring missions of those was NASA's Cosmic Background Explorer (COBE) satellite, which was launched to map large swathes of the sky in the 1990s.

The Big Bang theory is the leading scientific explanation of our universe's origins

DID YOU KNOW...?

It would be more accurate for the Big Bang theory to be called the "Everywhere Stretch"

Several other missions have followed in COBE's footsteps, such as the BOOMERanG experiment (Balloon Observations of Millimetric Extragalactic Radiation and Geophysics), NASA's Wilkinson Microwave Anisotropy Probe (WMAP) and the European Space Agency's Planck satellite.

Planck's observations, first released in 2013, mapped the background in unprecedented detail and revealed that the universe was older than previously thought: 13.82 billion years old, rather than 13.7 billion years old. The research observatory's mission is ongoing and new maps of the CMB are released periodically. The maps give rise to new mysteries, however, such as why the Southern Hemisphere appears slightly redder (warmer) than the Northern Hemisphere. The Big Bang theory says that the CMB would be mostly the same, no matter where you look.

Examining the CMB also gives astronomers clues as to the composition of the universe. Researchers think most of the cosmos is made up of matter and energy that cannot be "sensed" with conventional instruments, leading to the names 'dark matter' and 'dark energy'. Only 5 per cent of the universe is made up of matter such as planets, stars and galaxies, which we can see and sense.

GRAVITATIONAL WAVES CONTROVERSY

While astronomers could see the universe's beginnings, they've also been seeking out proof of its rapid inflation. Theory says that in the first second after the universe was born, our cosmos ballooned faster than the speed of light. That, by the way, does not violate Albert Einstein's speed limit since he said that light is the maximum anything can travel within the universe. That did not apply to the inflation of the universe itself.

In 2014, astronomers said they had found evidence in the CMB concerning "B-modes," a sort of polarisation generated as the universe got bigger and created gravitational waves. The team spotted evidence of this using an Antarctic telescope called "Background Imaging of Cosmic Extragalactic Polarization", or BICEP2.

"We're very confident that the signal that we're seeing is real, and it's on the sky," lead researcher John Kovac, of the Harvard-Smithsonian Center for Astrophysics, told Space.com in March 2014.

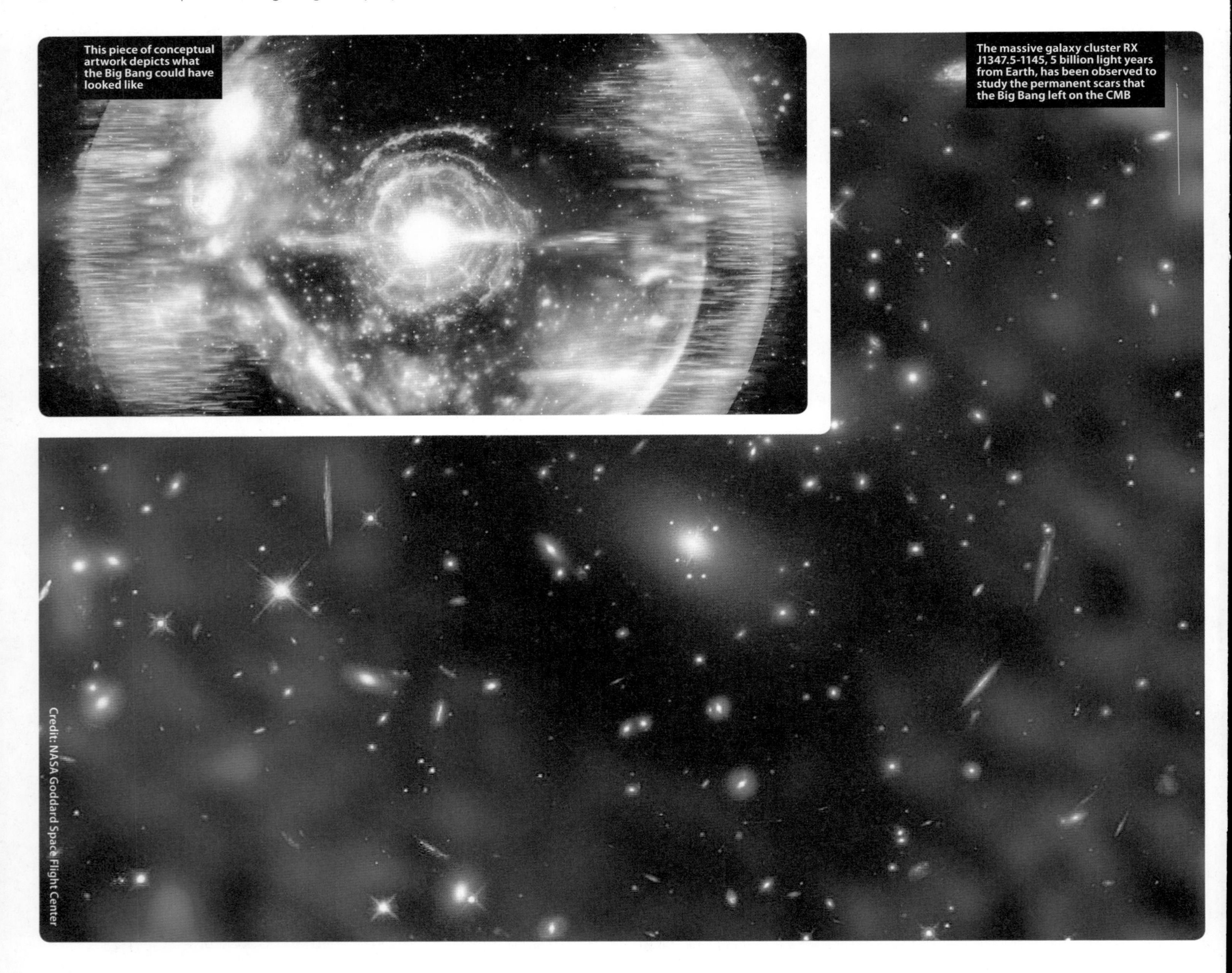

This piece of conceptual artwork depicts what the Big Bang could have looked like

The massive galaxy cluster RX J1347.5-1145, 5 billion light years from Earth, has been observed to study the permanent scars that the Big Bang left on the CMB

Credit: NASA Goddard Space Flight Center

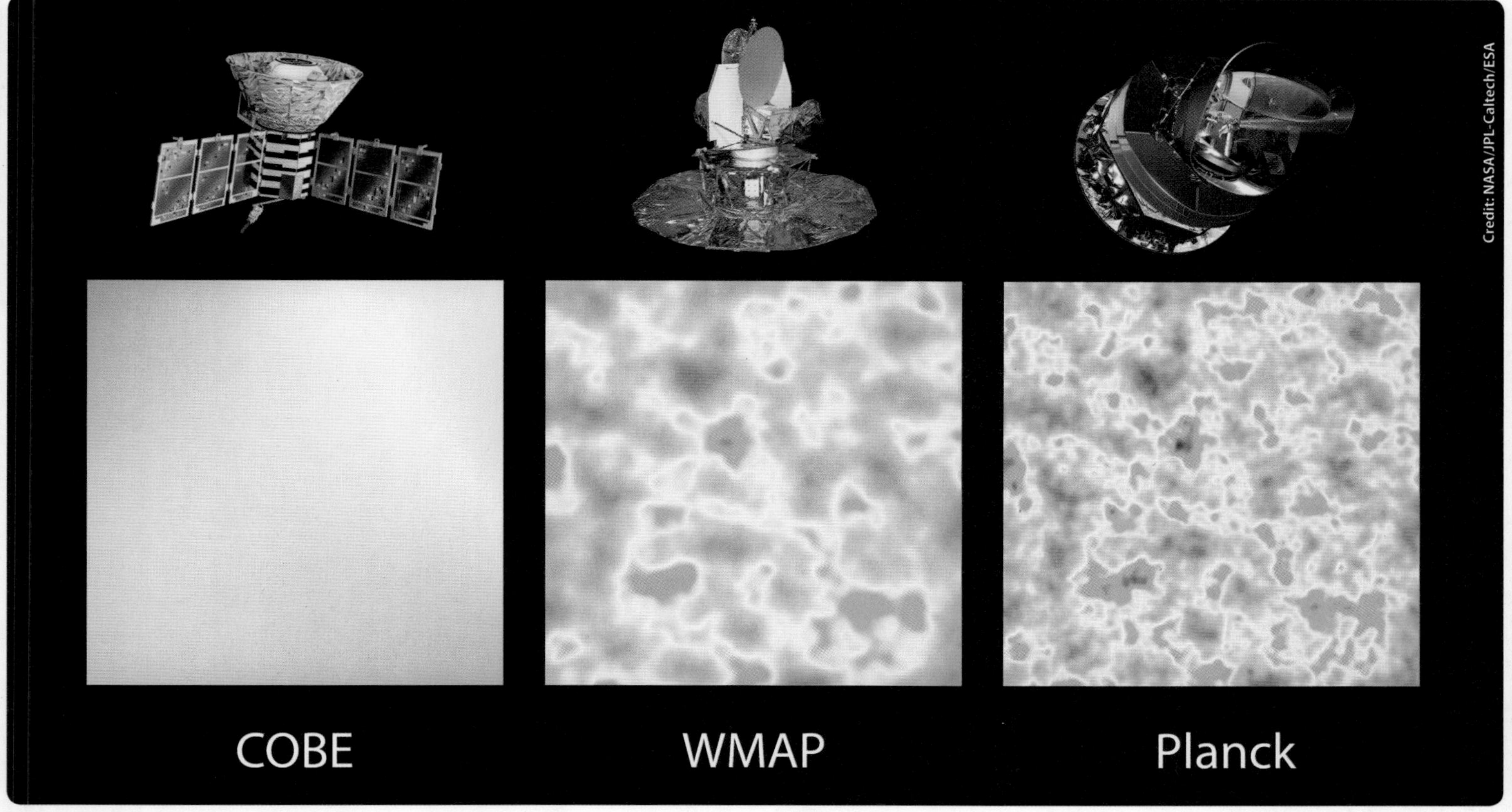

ABOVE: This graphic illustrates how successive satellites have provided us with increasingly detailed views of the ancient light left over from the Big Bang

But by June, the same team said that their findings could have been altered by galactic dust getting in the way of their field of view. "The basic takeaway has not changed; we have high confidence in our results," Kovac said in a press conference reported by the New York Times. "New information from Planck makes it look like pre-Planckian predictions of dust were too low," he added.

The results from Planck were put online in pre-published form in September. By January 2015, researchers from both teams working together "confirmed that the Bicep signal was mostly, if not all, stardust," the New York Times said in another article.

Separately, gravitational waves have been confirmed when talking about the movements and collisions of black holes that are a few tens of masses larger than our sun.

These waves have been detected multiple times by the Laser Interferometer Gravitational-Wave Observatory (LIGO) since 2016. As LIGO becomes more sensitive, it is anticipated that discovering black hole-related gravitational waves will be a fairly frequent event.

FASTER INFLATION, MULTIVERSES AND CHARTING THE START

The universe is not only expanding, but getting faster as it inflates. This means that with time, nobody will be able to spot other galaxies from Earth, or any other vantage point within our galaxy.

"We will see distant galaxies moving away from us, but their speed is increasing with time," Harvard University astronomer Avi Loeb said in a March 2014 Space.com article. "So, if you wait long enough, eventually, a distant galaxy will reach the speed of light. That means even light won't be able to bridge the gap being opened between that galaxy and us. There's no way for extraterrestrials on that galaxy to communicate with us, to send any signals that will reach us, once their galaxy is moving faster than light relative to us." Some physicists also suggest that the universe we experience is just one of many.

In the "multiverse" model, different universes would coexist with each other like bubbles lying side by side. The theory suggests that in that first big push of inflation, different parts of space-time grew at different rates.

This could have carved off different sections – different universes – with potentially different laws of physics. "It's hard to build models of inflation that don't lead to a multiverse," Alan Guth, a theoretical physicist at the Massachusetts Institute of Technology, said during a news conference in March 2014 concerning the gravitational waves discovery (Guth is not affiliated with that study). "It's not impossible, so I think there's still certainly research that needs to be done. But most models of inflation do lead to a multiverse, and evidence for inflation will be pushing us in the direction of taking [the idea of a] multiverse seriously."

While we can understand how the universe we see came to be, it's possible that the Big Bang was not the first inflationary period the universe experienced. Some scientists believe we live in a cosmos that goes through regular cycles of inflation and deflation, and that we just happen to be living in one of these phases.

"THE UNIVERSE IS NOT ONLY EXPANDING, BUT GETTING FASTER AS IT INFLATES. WITH TIME, NOBODY WILL BE ABLE TO SPOT OTHER GALAXIES FROM EARTH"

THE BIG BANG THEORY

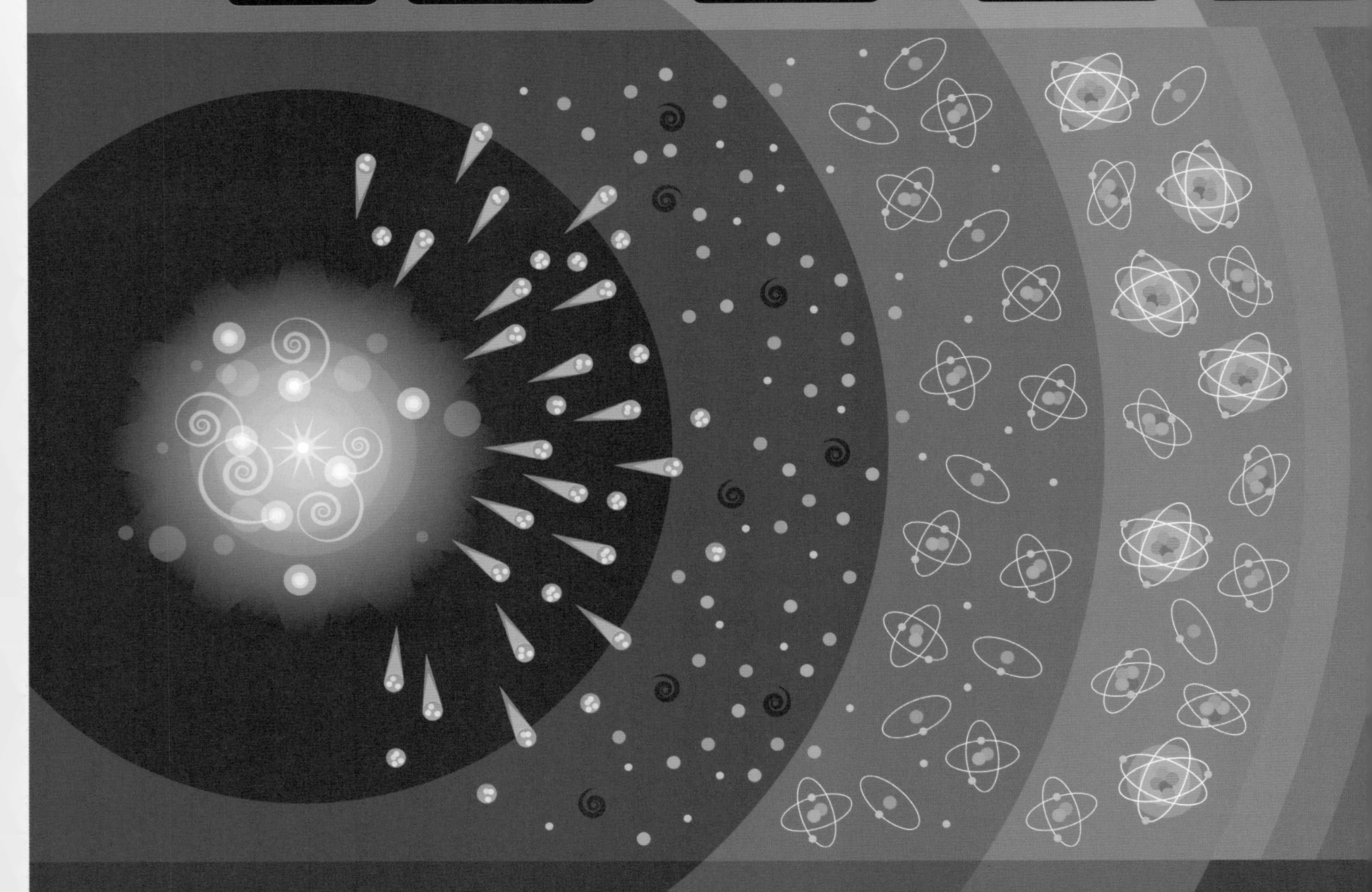

GRAVITY
STARS AND GALAXIES FORM

ANTIGRAVITY
UNIVERSE EXPANSION ACCELERATES

TODAY
UNIVERSE CONTINUES TO EXPAND

GALAXIES BREAK APART

300 MILLION YEARS

10 BILLION YEARS

13.8 BILLION YEARS

ALTERNATIVES TO THE BIG BANG THEORY EXPLAINED

MOST ASTRONOMERS BELIEVE THE UNIVERSE BEGAN 13.8 BILLION YEARS AGO IN A SUDDEN EXPLOSION CALLED THE BIG BANG. OTHER THEORISTS HAVE INVENTED ALTERNATIVES AND EXTENSIONS TO THIS THEORY

WORDS: KARL TATE

Nearly 14 billion years ago, there was nothing and nowhere. Then, all of a sudden, due to a random fluctuation in a completely empty void, a universe exploded into existence. Something the size of a subatomic particle inflated to an unimaginably huge size in a fraction of a second, driven apart by negative-pressure vacuum energy.

Scientists call this theory for the origin of the universe the Big Bang. What we call the "observable universe" (or the "Hubble Volume") is the spherical region, about 90 billion light-years in diameter, that is centered on any given observer. This is the only part of the universe in which light has had time to reach the observer in the 13.8 billion years since the universe began.

Since the universe's expansion is accelerating, objects are being dragged out of Earth's Hubble Volume and will become undetectable to humans of the future. The Hubble Volume is more than 13.8 billion light-years in radius because the expansion of space has increased distances between objects faster than light can travel.

ASTRONOMERS MAKE THREE ASSUMPTIONS ABOUT THE UNIVERSE BASED ON THEORY AND OBSERVATION:

- The laws of physics are universal and don't change with time or location in space.
- The universe is homogeneous, or roughly the same in every direction (though not necessarily for all of time).
- Humans do not observe the universe from a privileged location such as at its very centre.

WHEN THESE ASSUMPTIONS ARE APPLIED TO ALBERT EINSTEIN'S EQUATIONS, THEY INDICATE THE UNIVERSE HAS THE FOLLOWING PROPERTIES:

- The universe expands (astronomers see light from the universe's distant regions shifted toward the red end of the spectrum by the expansion of the space between).
- The universe emerged from a hot, dense state at some finite time in the past.
- The lightest elements, hydrogen and helium, were created in the first moments of time.
- A background of microwave radiation fills the entire universe, a relic of the phase transition that occurred when the hot, early universe cooled enough for atoms to form.

If any of astronomers' basic assumptions are wrong, the Big Bang theory would not explain the properties of this universe.

IS IT POSSIBLE THAT THE BIG BANG NEVER HAPPENED AT ALL?

One alternative theory is the Steady State universe. An early rival to the Big Bang theory, Steady State posits continuous creation of matter throughout the universe to explain its apparent expansion. This type of universe would be infinite, with no beginning or end. However, a mountain of evidence found since the mid-1960s indicates this theory is not correct.

Another alternative is the Eternal Inflation theory. After the Big Bang, the universe expanded rapidly during a brief period called inflation. The Eternal Inflation theory posits that inflation never stopped, and has been going on for an infinite length of time. Somewhere, even now, new universes are continuously coming into existence in a vast complex called the multiverse.

Those many universes could have different physical laws. The Oscillating Model of the Universe involved an endless series of Big Bangs, followed by Big Crunches that restarted the cycle, endlessly. The modern cyclic model involves colliding "branes" (a "membrane" within a higher-dimensional volume called the "bulk").

Implications found in quantum gravity and string theory tantalisingly suggest a universe that is in reality nothing like how it appears to human observers. It may actually be a flat hologram projected onto the surface of a sphere, for example. Or it could be a completely digital simulation running in a vast computer.

DID YOU KNOW...?

One of the problems with the Big Bang theory is that it seems to defy the first law of thermodynamics

According to the Simulation hypothesis, the entire universe could just be an artificial simulation

COSMIC MICROWAVE BACKGROUND

WHAT CAN THE REMNANTS OF THE BIG BANG TELL US ABOUT THE AGE OF THE UNIVERSE?

WORDS: ELIZABETH HOWELL

The cosmic microwave background (CMB) is thought to be leftover radiation from the Big Bang, or the time when the universe began. As the theory goes, when the universe was born it underwent a rapid inflation and expansion. The universe is still expanding today, and the expansion rate appears different depending on where you look. The CMB represents the heat left over from the Big Bang.

You can't see the CMB with your naked eye, but it is everywhere in the universe. It is invisible to humans because it is so cold, just 2.725 degrees above absolute zero (minus 459.67 degrees Fahrenheit, or minus 273.15 degrees Celsius.) This means its radiation is most visible in the microwave part of the electromagnetic spectrum.

ORIGINS AND DISCOVERY

The universe began 13.8 billion years ago, and the CMB dates back to about 400,000 years after the Big Bang. That's because in the early stages of the universe, when it was just one-hundred-millionth the size it is today, its temperature was extreme: 273 million degrees above absolute zero.

Any atoms present at that time were quickly broken apart into small particles (protons and electrons). The radiation from the CMB in photons (particles representing quantums of light, or other radiation) was scattered off the electrons. "Thus, photons wandered through the early universe, just as optical light wanders through a dense fog," explained NASA.

About 380,000 years after the Big Bang, the universe was cool enough that hydrogen could form. Because the CMB photons are barely affected by hitting hydrogen, the photons travel in straight lines. Cosmologists refer to a "surface of last scattering" when the CMB photons last hit matter; after that, the universe was too big. So when we map the cosmic microwave background, we are looking back in time to 380,000 years after the Big Bang, just after the universe was opaque to radiation.

American cosmologist Ralph Apher first predicted the CMB in 1948, when he was doing work with Robert Herman and George Gamow, according to NASA. The team was doing research related to Big Bang nucleosynthesis, or the production of elements in the universe besides the lightest isotope (type) of hydrogen. This type of hydrogen was created very early in the universe's history.

But the CMB was first found by accident. In 1965, two researchers with Bell Telephone Laboratories (Arno Penzias and Robert Wilson) were creating a radio receiver, and were puzzled by the noise it was picking up. They soon realized the noise came uniformly from all over the sky. At the same time, a team at Princeton University (led by Robert Dicke) was trying to find the CMB. Dicke's team got wind of the Bell experiment and realized the CMB had been found. Both teams quickly published papers in the Astrophysical Journal in 1965, with Penzias and Wilson talking about what they saw, and Dicke's team explaining what it means in the context of the

An artist's impression of star formation in the early universe

Credit: Adolf Schaller for STScI

The Planck mission provided a more detailed view of the cosmic microwave background

Credit: ESA and the Planck Collaboration

"THIS 'BABY PICTURE' OF THE UNIVERSE CONFIRMED BIG BANG THEORY PREDICTIONS"

Robert Wilson (left) and Arno Penzias (right) in front of the antenna with which they discovered CMB

An artist's impression of the Planck spacecraft, which launched in 2009

Credit: ESA

universe. (Later, Penzias and Wilson both received the 1978 Nobel Prize in physics).

STUDYING IN MORE DETAIL

The CMB is useful to scientists because it helps us learn how the early universe was formed. It is at a uniform temperature with only small fluctuations visible with precise telescopes. "By studying these fluctuations, cosmologists can learn about the origin of galaxies and large-scale structures of galaxies and they can measure the basic parameters of the Big Bang theory," NASA wrote.

While portions of the CMB were mapped in the ensuing decades after its discovery, the first space-based full-sky map came from NASA's Cosmic Background Explorer (COBE) mission, which launched in 1989 and ceased science operations in 1993. This "baby picture" of the universe, as NASA calls it, confirmed Big Bang theory predictions and also showed hints of cosmic structure that were not seen before. In 2006, the Nobel Prize in physics was awarded to COBE scientists John Mather at the NASA Goddard Space Flight Center, and George Smoot at the University of California, Berkeley.

A more detailed map came in 2003 courtesy of the Wilkinson Microwave Anisotropy Probe (WMAP), which launched in June 2001 and stopped collecting science data in 2010. The first picture pegged the universe's age at 13.7 billion years (a measurement that has since been refined to 13.8 billion years) and also revealed a surprise: the oldest stars started shining about 200 million years after the Big Bang, far earlier than predicted.

Scientists followed up those results by studying the very early inflation stages of the universe (in the trillionth second after formation) and by giving more precise parameters on atom density, the universe's lumpiness and other properties of the universe shortly after it was formed. They also saw a strange asymmetry in average temperatures in both hemispheres of the sky, and a "cold spot" that was bigger than expected. The WMAP team received the 2018 Breakthrough Prize in Fundamental Physics for their work.

In 2013, data from the European Space Agency's Planck space telescope was released, showing the highest precision picture of the CMB yet. Scientists uncovered another mystery with this information: Fluctuations in the CMB at large angular scales did not match predictions. Planck also confirmed what WMAP saw in terms of the asymmetry and the cold spot. Planck's final data release in 2018 (the mission operated between 2009 and 2013) showed more proof that dark matter and dark energy – mysterious forces that are likely behind the acceleration of the universe – do seem to exist.

THE FIRST STARS

HOW THE 'COSMIC DAWN' BROKE AND THE FIRST STARS FORMED

WORDS: PAUL SUTTER

Perhaps the greatest revelation in the past hundred years of studying the universe is that our home changes and evolves with time. And not just in minor, insignificant ways like stars moving about, gas clouds compressing and massive stars dying in cataclysmic explosions. No, our entire cosmos has changed its fundamental character more than once in the distant past, completely altering its internal state at a global – that is, universal – scale. Take, for instance, the fact that at one time in the foggy, ill-remembered past, there were no stars.

BEFORE THE FIRST LIGHT

We know this simple fact because of the existence of the cosmic microwave background (CMB), a bath of weak but persistent radiation that soaks the entire universe. If you encounter a random photon (a bit of light), there's a good chance it's from the CMB – that light takes up more than 99.99 percent of all the radiation in the universe. It's a leftover relic from when the universe was just 270,000 years old, and transitioned from a hot, roiling plasma into a neutral soup (with no positive or negative charge). That transition released white-hot radiation that, over the course of 13.8 billion years, cooled and stretched down into the microwaves, giving us the background light that we can detect today.

At the time of the release of the CMB, the universe was about one-millionth its present volume and thousands of degrees hotter. It was also almost entirely uniform, with density differences no bigger than 1 part in 100,000. So, not exactly a state where stars could happily exist.

THE DARK AGES

In the millions of years following the release of the CMB (known as "recombination" in astronomy circles, due to a historical misunderstanding of even earlier epochs), the universe was in an odd state.

There was a still a persistent bath of white-hot radiation, but that radiation was quickly cooling as the universe continued its inexorable expansion. There was dark matter, of course, hanging out minding its own business.

And there was also the now-neutral gas, almost entirely hydrogen and helium at this point, finally released from its struggles with radiation and free to do as it pleased.

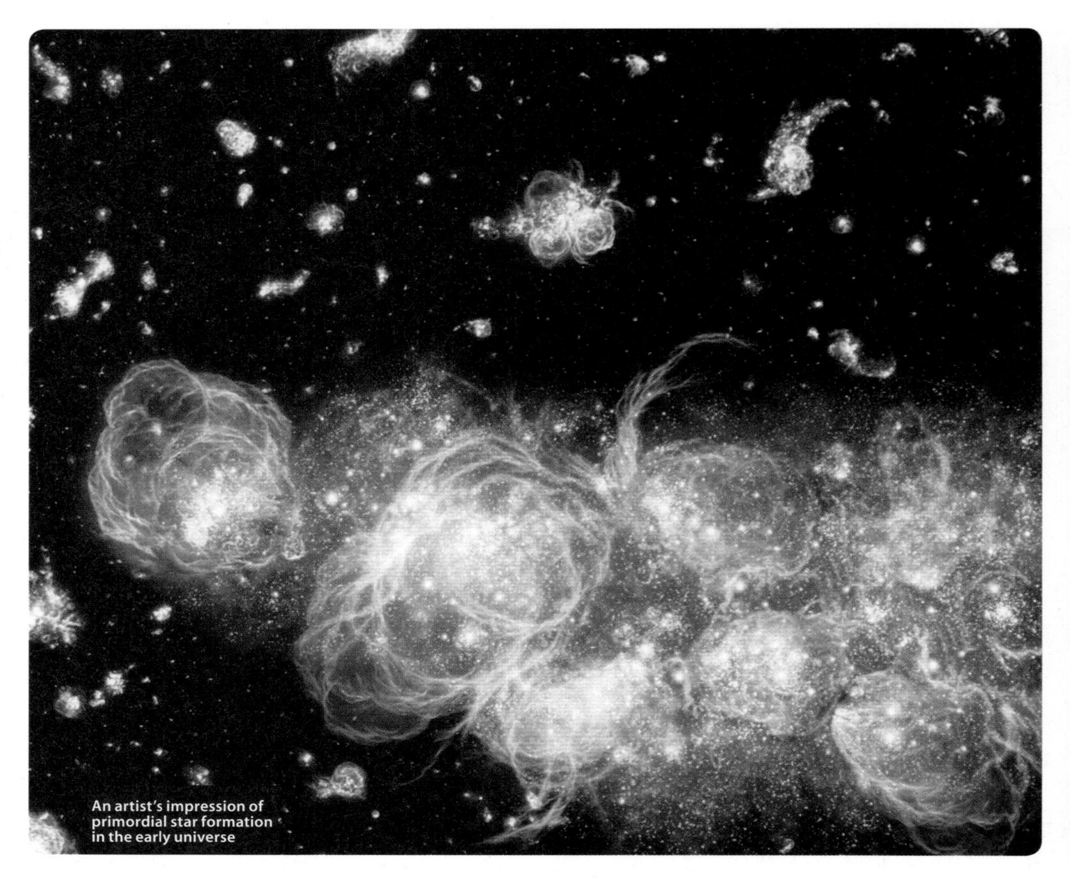

An artist's impression of primordial star formation in the early universe

And what it pleased to do was to hang out with as much of itself as possible. Thankfully, it didn't have to work very hard: In the exceedingly early universe, microscopic quantum fluctuations enlarged to become merely small differences in density (and why that happened is a story for another day).

These tiny density differences didn't affect the greater cosmological expansion, but they did impact the lives of that neutral hydrogen. Any one patch that was slightly denser than average – even by a tiny, tiny bit – had a slightly stronger gravitational pull on its neighbors. That enhanced pull encouraged more gas to join the party, which amplified the gravitational tug, which encouraged even more neighbors and so on.

Like loud music at a party acting as a siren song to attract more revelers, over the course of millions of years the rich gas got richer and the poor gas got poorer. Through simple gravity, tiny density differences grew, building the first massive blocks of matter and emptying out their surroundings.

THE "COSMIC DAWN" BREAKS

Somewhere, someplace, some chunk of neutral hydrogen got lucky. Piling layers upon overwhelming layers on itself, the innermost core reached a critical temperature and density, forcing the atomic nuclei together in a complicated pattern, igniting in nuclear fusion and converting the raw material into helium. That ferocious process also released a little bit of energy, and in a flash the first star was born.

For the first time since the first dozen minutes of the Big Bang, nuclear reactions took place in our universe. New sources of light, dotting the cosmos, flooded the once-empty voids with radiation. But we're not exactly sure when this momentous event occurred; observations of this epoch are exceedingly difficult. For one, the vast cosmological distances

"FOR THE FIRST TIME SINCE THE FIRST DOZEN MINUTES OF THE BIG BANG, NUCLEAR REACTIONS TOOK PLACE IN OUR UNIVERSE... NEW SOURCES OF LIGHT FLOODED THE ONCE-EMPTY VOIDS WITH RADIATION"

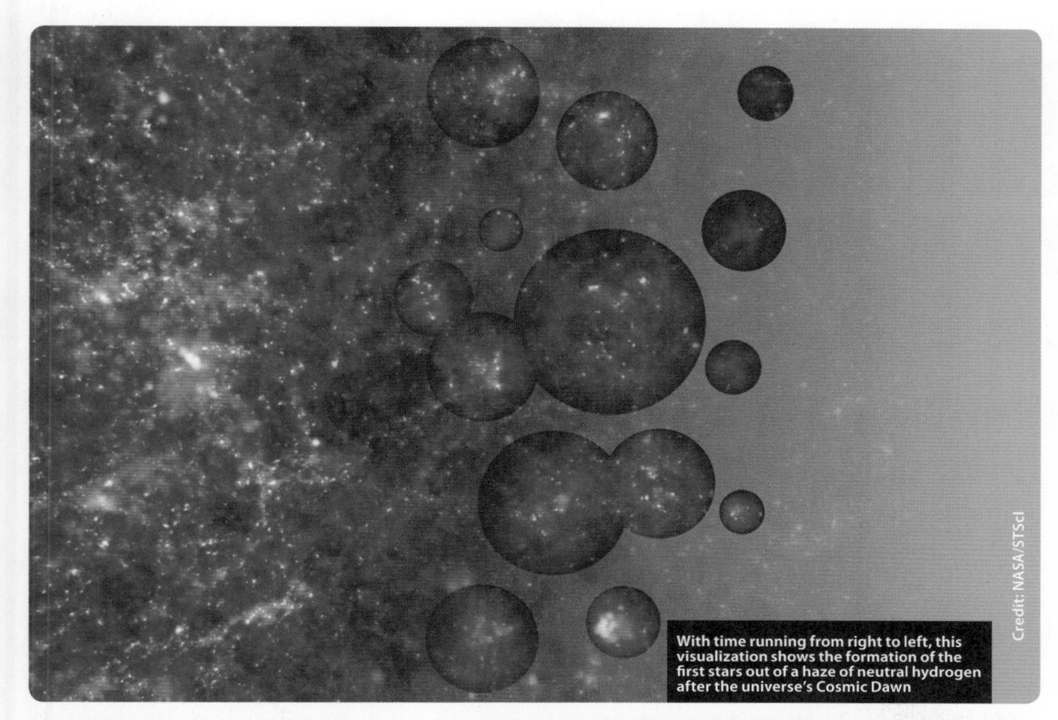

With time running from right to left, this visualization shows the formation of the first stars out of a haze of neutral hydrogen after the universe's Cosmic Dawn

Credit: NASA/STScI

An artist's impression of the distant galaxy CR7, which has provided astronomers with strong evidence of the first generation of stars

Credit: ESO/M. Kornmesser

prevent even our most powerful telescopes from observing that first light. What makes it worse is that the early universe was almost entirely neutral, and neutral gas doesn't emit a lot of light in the first place. It's not until multiple generations of stars glue themselves together to form galaxies that we can even get a dim hint of this important age.

We suspect that the first stars formed somewhere within the first few hundred million years of the universe. It's not much later that we have direct observations of galaxies, active galactic nuclei and even the beginnings of clusters of galaxies – the most massive structures to eventually arise in the universe. Sometime before them the first stars had to arrive, but not too early, because the hectic conditions of the infant universe would have prevented their formation.

OVER THE HORIZON

Although the upcoming James Webb Space Telescope will be able pinpoint early galaxies with excellent precision, offering a wealth of data on the early universe, the telescope's narrow field of view won't give us the whole picture of this era. Scientists hope that some of the earliest galaxies might contain remnants of the very first stars – or even the stars themselves – but we'll have to wait and (literally) see.

The other way to unlock the cosmic dawn is through a surprising quirk of neutral hydrogen. When the quantum spins of the electron and proton randomly flip, the hydrogen emits radiation of a very specific wavelength: 21 centimeters. This radiation allows us to map out pockets of neutral hydrogen in our modern-day Milky Way, but the extreme distances to the cosmic dawn era pose a different challenge altogether. The trouble is that the universe has expanded since that long-dead era, which causes all intergalactic radiation to stretch out to longer wavelengths. Nowadays, that primordial neutral hydrogen signal has a wavelength of around 2 meters, placing the signal firmly in the radio bands. And many other things in the universe – supernovas, galactic magnetic fields, satellites – are quite loud at those same frequencies, obscuring the faint signal from the universe's early years.

There are several missions around the globe trying to home in on that juicy cosmic- dawn signal, dig out its primordial whisper from the present-day cacophony, and reveal the birth of the first stars. But for now, we'll just have to wait and listen.

PARALLEL UNIVERSES: THEORIES & EVIDENCE

OUR UNIVERSE MAY LIVE IN ONE BUBBLE THAT IS SITTING IN A NETWORK OF BUBBLE UNIVERSES IN SPACE

WORDS: ELIZABETH HOWELL

Is our universe unique? From science fiction to science fact, there is a concept that suggests that there could be other universes besides our own, where all the choices you made in this life played out in alternate realities. The concept is known as a "parallel universe," and is a facet of the astronomical theory of the multiverse.

The idea is pervasive in comic books, video games, television and movies. Franchises ranging from "Buffy the Vampire Slayer" to "Star Trek" and "Doctor Who" to "Digimon" use the idea to extend plotlines (see the boxout on page 33 for more pop culture examples). There actually is quite a bit of evidence out there for a multiverse. First, it is useful to understand how our universe is believed to have come to be.

DID YOU KNOW...?

An idea from string theory is that parallel 'braneworlds' exist in a higher-dimensional space

WHAT ARE THE ARGUMENTS FOR A MULTIVERSE?

Around 13.8 billion years ago, simply speaking, everything we know of in the cosmos was an infinitesimal singularity. Then, according to the Big Bang theory, some unknown trigger caused it to expand and inflate in three-dimensional space. As the immense energy of this initial expansion cooled, light began to shine through. Eventually, the small particles began to form into the larger pieces of matter we know today, such as galaxies, stars and planets. One big question with this theory is: are we the only universe out there? With our current technology, we are limited to observations within this universe because the universe is curved and we are inside the fishbowl, unable to see the outside of it (if there is an outside). There are at least five main theories why a multiverse is possible:

1. INFINITE UNIVERSES

We don't know what the shape of space-time is exactly. One prominent theory is that it is flat and goes on forever. This would present the possibility of many universes being out there. But with that topic in mind, it's possible that universes can start repeating themselves. That's because particles can only be put together in so many ways. More about that in a moment.

2. BUBBLE UNIVERSES

Another theory for multiple universes comes from "eternal inflation." Based on research from Tufts University cosmologist Alexander Vilenkin, when

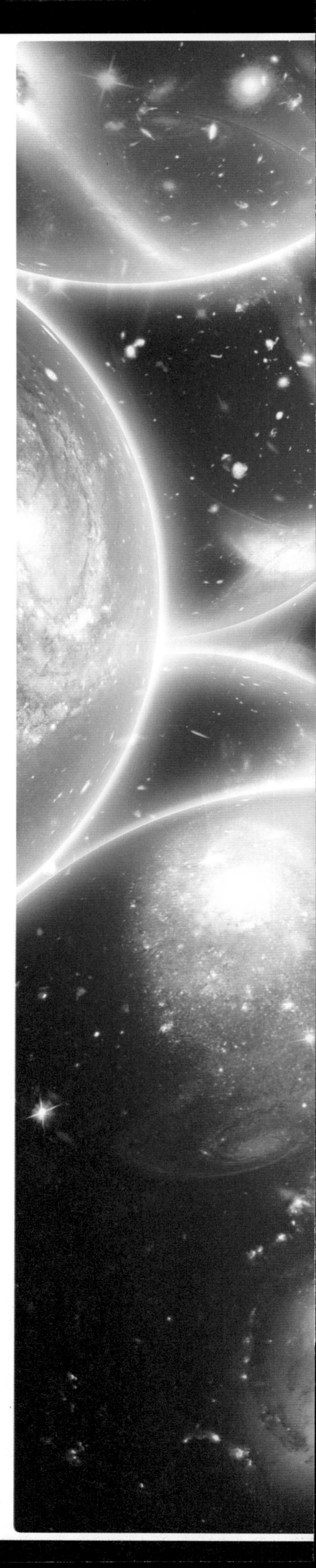

Our universe could just be one of an infinite number of others

Researchers from Durham University in the UK posit that the anomalous 'Cold Spot' on the cosmic microwave background radiation could be the result of a parallel universe crashing into our own

looking at space-time as a whole, some areas of space stop inflating like the Big Bang inflated our own universe. Others, however, will keep getting larger. So if we picture our own universe as a bubble, it is sitting in a network of other bubble universes of space. What's interesting about this theory is the other universes could have very different laws of physics than our own, since they are not linked.

3. DAUGHTER UNIVERSES

Or perhaps multiple universes can follow the theory of quantum mechanics (how subatomic particles behave), as part of the "daughter universe" theory. If you follow the laws of probability, it suggests that for every outcome that could come from one of your decisions, there would be a range of universes – each of which saw one outcome come to be. So in one universe, you took that job in China. In another, perhaps you were on your way and your plane landed somewhere different, and you decided to stay. And so on.

4. MATHEMATICAL UNIVERSES

Another possible avenue is exploring mathematical universes, which, simply put, explain that the structure of mathematics may change depending in which universe you reside. "A mathematical structure is something that you can describe in a way that's completely independent of human baggage," said theory-proposer Max Tegmark of MIT as quoted in the 2012 article. "I really believe that there is this universe out there that can exist independently of me that would continue to exist even if there were no humans."

5. PARALLEL UNIVERSES

And last but not least is the idea of parallel universes. Going back to the idea that space-time is flat, the number of possible particle configurations in multiple universes would be limited to 10^10^122 distinct possibilities, to be exact. So, with an infinite number of cosmic patches, the particle arrangements within them must repeat – infinitely many times over. This means there are infinitely many "parallel universes": cosmic patches exactly the same as ours (containing someone exactly like you), as well as patches that differ by just one particle's position, patches that differ by two particles' positions, and so on, down to patches that are totally different from ours.

Renowned theoretical physicist and cosmologist Stephen Hawking's final paper also dealt with the idea of a multiverse. The paper was published in May 2018, just a few months after Hawking's death. Talking about the theory, he explained to Cambridge University in an interview published in The Washington Post, "We are not down to a single, unique universe, but our findings imply a significant reduction of the multiverse to a much smaller range of possible universes."

THE ARGUMENTS AGAINST PARALLEL UNIVERSES

Not everyone agrees with the parallel universe theory, however. A 2015 article on Medium by astrophysicist Ethan Siegal agreed that space-time could go on forever in theory, but said that there are some limitations with that idea.

The key problem is the universe is just under 14 billion years old. So our universe's age itself is obviously not infinite, but a finite amount. This would (simply put) limit the number of possibilities for particles to rearrange themselves, and sadly make it less possible that your alternate self did get on that plane to China after all.

Also, the expansion at the beginning of the universe took place exponentially because there was so much "energy inherent to space itself," he said. But over time, that inflation obviously slowed – those particles of matter created at the Big Bang are not continuing to expand, he pointed out.

Among his conclusions: that means that multiverses would have different rates of inflation and different times (longer or shorter) for inflation. This decreases the possibilities of universes similar to our own. "Even setting aside issues that there may be an infinite number of possible values for fundamental constants, particles and interactions, and even setting aside interpretation issues such as whether the many-worlds-interpretation actually describes our physical reality," Siegal said, "the fact of the matter is that the number of possible outcomes rises so quickly – so much faster than merely exponentially – that unless inflation has been occurring for a truly infinite amount of time, there are no parallel universes identical to this one."

But rather than seeing this lack of other universes as a limitation, Siegal instead takes the philosophy that it shows how important it is to celebrate being unique. He advises to make the choices that work for you, which "leave you with no regrets." That's because there are no other realities where the choices of your dream self play out; you are the only person that can make those choices happen.

"RENOWNED PHYSICIST AND COSMOLOGIST STEPHEN HAWKING'S FINAL PAPER ALSO DEALT WITH THE IDEA OF A MULTIVERSE"

PARALLEL UNIVERSES IN SCIENCE FICTION

HERE ARE JUST SOME OF THE MANY USES OF PARALLEL UNIVERSES IN POPULAR CULTURE

- Marvel Comics and DC Comics feature stories set in parallel universes that are part of the multiverse.
- Many anime series, such as "Digimon," "Dragon Ball" and "Sonic the Hedgehog" feature alternate versions of their characters from other universes.
- Parallel universes appear in many video games, including "Dungeons & Dragons," "BioShock Infinite," the "Final Fantasy" franchise, "Half-Life," "League of Legends," "Mortal Kombat" and "The Legend of Zelda."
- "Flatland: A Romance of Many Dimensions" (1884), by Edwin A. Abbott, is a story about a two-dimensional world that includes living geometric figures such as circles, triangles and squares. The novel also includes other universes such as Lineland, Spaceland and Pointland. This book was adapted into a feature film in 2007.
- "Men Like Gods" (1923), an H.G. Wells novel, included a "paratime" machine and explored the idea of a multiverse.
- "The Chronicles of Narnia" (1950-56), a C.S. Lewis book series, features several children who move between our world and the world of Narnia, where there are talking animals. Some of these books were released as feature films earlier in the 2000s.
- An episode of "Star Trek" featured a "mirror universe" in which the characters were more ruthless and warlike. The concept was repeated in nearly every subsequent "Star Trek" series. In 2009, the "Star Trek" universe got a reboot in a movie that put the characters from the 1960s original series in an alternate universe. The movie starred Chris Pine and Zachary Quinto and set off a series of other "Star Trek" films.
- In "The Dark Tower," a Stephen King series that began in 1982, travellers go through portals to different levels of the titular tower (in other words, parallel Earths). Part of the series was adapted into a feature film in 2017.
- The "Back to the Future" movie series (which began in 1985) follows the adventures of the McFly family, including visits to 1885, 1955 and 2015. The second film in particular shows the drawbacks of an alternate reality, when one character uses it to get rich by nefarious means. The series starred Michael J. Fox.
- In the "His Dark Materials" series by Philip Pullman, children move between multiple worlds. The first book, "The Golden Compass," was adapted into a film in 2007.
- "Sliding Doors" (1998) is a film that showed two parallel universes depending on whether the main character caught a train or not. It starred Gwyneth Paltrow and John Hannah.
- "Run Lola Run" (1998) is a film starring Franka Potente. The film shows multiple alternatives as a woman tries to get 100,000 Deutsche marks in only 20 minutes to save her boyfriend's life.
- "Timeline" (1999) by Michael Crichton follows historians who go back in time to the Middle Ages. (While the book is mostly a time travel book, the multiverse is used in it as well.) A film based on the book was released in 2003.
- "Donnie Darko" (2001) is a film in which a high school student (played by Jake Gyllenhaal) finds himself confronted with visions and tries to figure out their meaning.
- "The Long Earth" book series, by Terry Pratchett and Stephen Baxter, discusses parallel universes that may be nearly the same as Earth.
- The Netflix sci-fi horror series "Stranger Things" (2016 to present) features an alternate dimension – which the characters refer to as "the Upside Down" – that starts to affect their own reality.

"The Upside Down" in Netflix's "Stranger Things'" is a nightmarish alternate universe

14 NUMBERS OF THE UNIVERSE

OUR UNIVERSE IS TEEMING WITH MIND-BENDINGLY BIG NUMBERS

WORDS: YASEMIN SAPLAKOGLU

From all the grains of sand on Earth to the number of stars in the sky, With the addition of only a few zeros, big numbers transform from "countable" to the subject of guesswork. Eventually, their existence pulls at our imagination and requires the crafting of intricate scenarios. They may or may not have a presence in the universe. From the tiniest specks in the universe to the biggest numbers humanity has ever conceived, here are some of the numbers that make up our universe.

1 ZERO

The total energy that makes up the microbes, plants, oceans, planets, stars and galaxies (in other words, our entire universe) is probably... zero. That's because the negative energy in the universe most likely cancels out the positive energy. Physicists consider light, matter and antimatter to be positive energy, while all the gravitational energy between particles has a negative charge. So, everything balances out.

2 HALF A MILLION

There are over half a million pieces of space junk – both meteors and artificial particles – larger than the size of a marble that orbit the Earth. Millions more are too small to track. This computer-generated image (right) illustrates space junk in the geosynchronous region, around 22,235 miles (35,785 kilometres) in altitude above the Earth's equator.

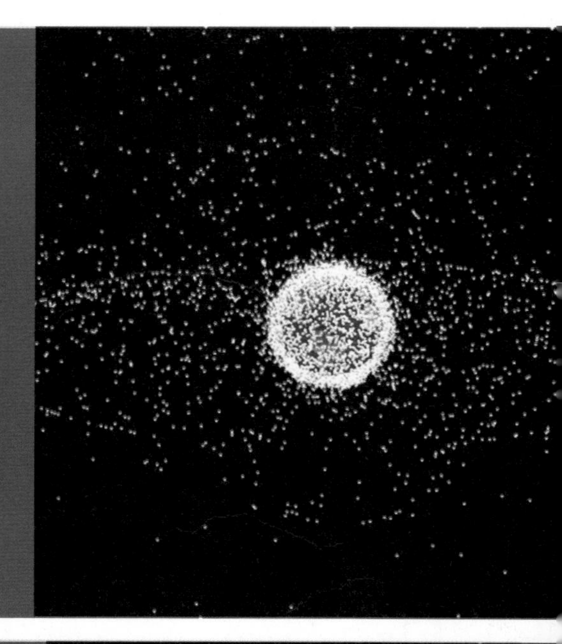

3 1 MILLION

One million planets, all capable of supporting life, could theoretically orbit a supermassive black hole. Astrophysicist Sean Raymond calculated that a black hole that has a mass 1 million times that of the sun and a ring of nine sun-like stars around it could hold 400 rings of planets. Each ring would have 2,500 Earth-mass planets. In such galaxies, "You would never feel alone," Raymond said. "Other planets would loom huge in the sky." This is just one scenario of possible planetary systems – and a very cramped one at that.

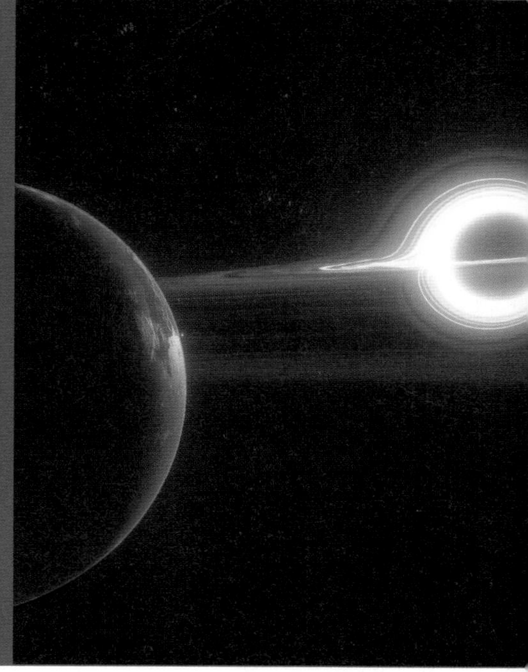

4 3 TRILLION

According to recent assessments, there are over 3 trillion trees in the world. But this is just an estimate, and the true number could change. While this is much larger than previous best estimates (which were around 400 billion) each year, humans may be removing around 15 billion trees while planting just 5 billion. Since the last ice age, approximately 11,000 years ago, humans may have already removed 3 trillion trees, according to the BBC.

5 QUADRILLION

The Earth's interior may be filled with a quadrillion tons of diamonds. But these diamonds are unreachable, located around 90 to 150 miles (145 to 240 kilometres) below the surface of the Earth in "roots" of cratons, or large sections of rock that lie beneath most continental tectonic plates. A group of scientists found that seismic waves, or vibrations that run below Earth's surface and vary based on the makeup of the rocks they travel through, tended to speed up when moving through cratonic roots. Their speeds matched those found for virtual rock models composed in part by diamonds.

6 QUINTILLION

Have you ever wanted to count the grains of sand on a beach? Scientists estimated that there are around 7 quintillion grains of sand on all the world's beaches, according to NPR. Well, actually 7 quintillion, 500 quadrillion grains of sand, or 7.5 with 18 zeroes after it. Now, the question is this: To prove that experimentally, would we ever have time to count it all?

7 SEXTILLION

Humanity may have left 24 sextillion, 640 quintillion (24.64x10^21) footprints on this planet since our very early days, according to Newsweek. This calculation was made assuming an average person walks 10,000 steps a day and lives to the age of 65. In a March study in the journal PLOS One, scientists described some of the oldest human footprints found in North America, dating back 13,000 years.

Credit: NASA; Getty; Thinkstock; Wiki/ Pablo Carlos Budassi (Creative Commons 3.0)

8 SEPTILLION

There are around 1 septillion (10^24) stars in the universe. This calculation assumed there were around 10 trillion galaxies in the universe and multiplied that by the Milky Way's estimated 100 billion stars. But even this giant number could be an underestimate since we don't really know how large the universe is. The observable universe goes back around 13.8 billion years. Beyond that, it could be infinite.

9 OCTILLION

There are around 920 to 3,170 octillion microbes on the planet (920 x 10^27 to 3170 x 10^27), Steven D'Hondt, a professor of oceanography at the University of Rhode Island explained.

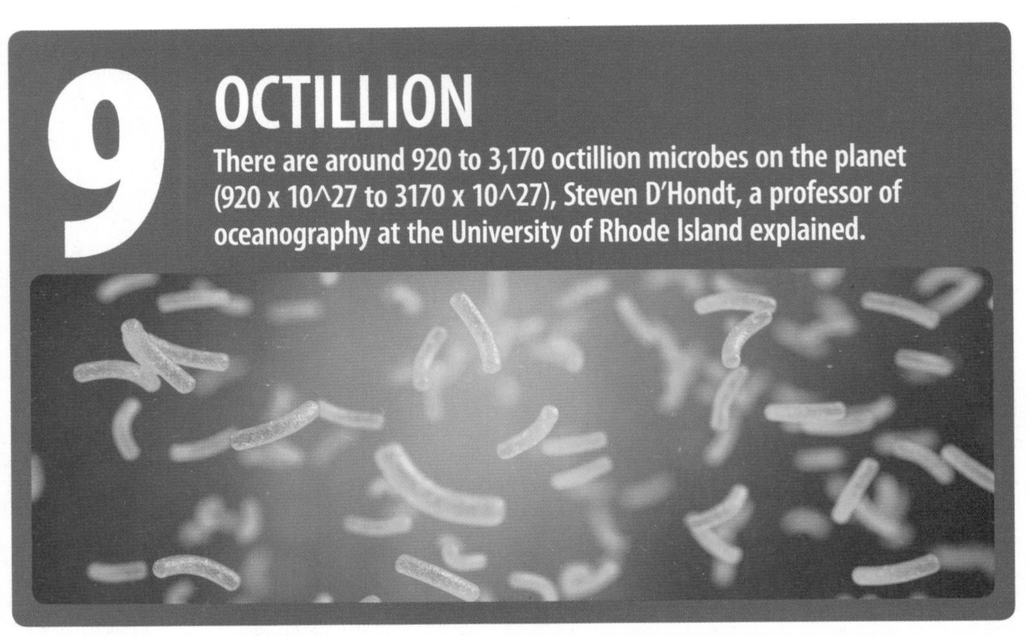

10 NONILLION

It would take around 160 nonillion (160 x 10^30) Great Pyramids of Giza to match the mass of the Milky Way, according to Inverse.

11 SEXDECILLION

The mass of the observable universe is 30 sexdecillion kilograms (30 x 10^51 kg), which is equivalent to about 25 billion galaxies the size of the Milky Way, according to astronomer Jagadheep D. Pandian, who answered the question on Cornell University's "Ask an Astronomer" page.

12 QUINVIGINTILLION

There are around 100 quinvigintillion atoms on the planet, or 100x10^78. By mass, around 75% of the universe is hydrogen and 25% is helium.

13 GOOGOLPLEX

If you fill the entire observable universe with fine dust particles around 1.5 micrometers big, the total number of combinations in which these particles can be arranged is equal to one googolplex, according to astronomer and astrophysicist Carl Sagan. But others have a different idea about what a googolplex means.

American mathematician Edward Kasner invented the number googol to describe 10 to the hundredth power (10^100). But he actually credited the name to his nephew, Milton Sirotta, who coined it in 1920 at the age of 9. Milton then came up with the number "googolplex", which he said should be "one, followed by writing zeros until you get tired."

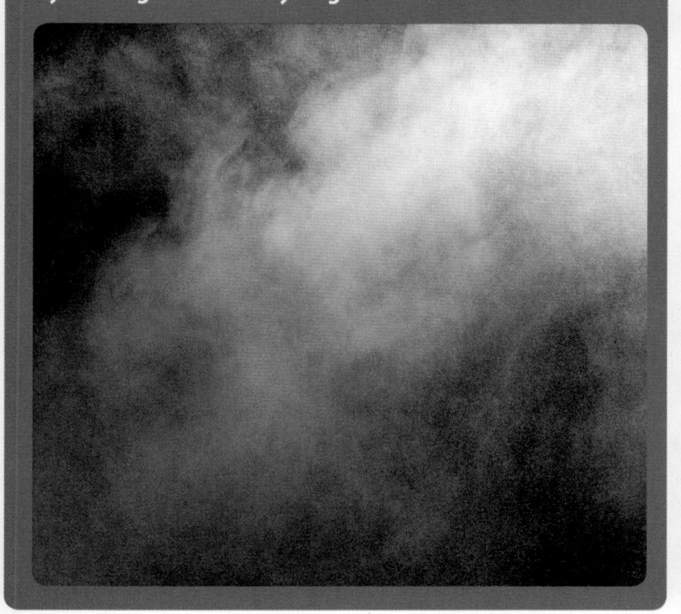

14 THE NUMBER SO BIG IT HAS NO NAME

Take a collection of books that are each 410 pages long, with 3,200 characters per page; the number of these books that would hold every combination of characters (i.e., every possible sequence of every book that has or possibly could be written in any language and even gibberish) would be around 10 to the power of 2 million, or 10 followed by 2 million zeros, according to Smithsonian magazine. This is "The Total Library," imagined by Argentine writer Jorge Luis Borges. The calculations were made by Jonathan Basile, who studied English literature at Columbia University and decided to create a digital version of Borges' library, according to Smithsonian.

Credit: NASA; Getty; Thinkstock; Wiki/ Pablo Carlos Budassi (Creative Commons 3.0)

THE SOLAR SYSTEM

Credit: NASA/GSFC

Credit: NASA

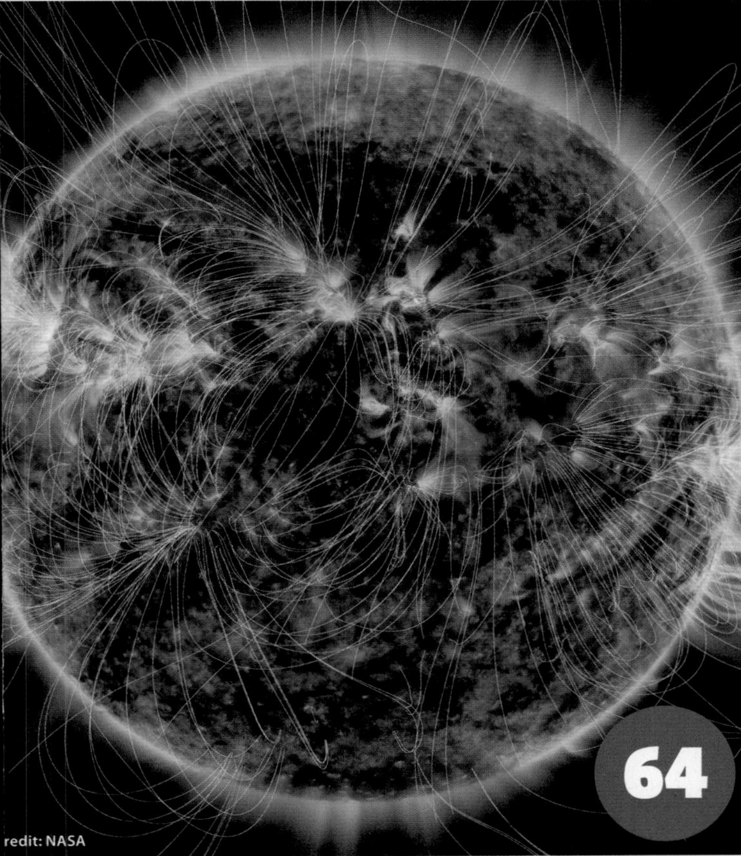

redit: NASA

Credit: Getty Images

THE PLANETS

WHILE MANY PEOPLE CAN POINT TO A PICTURE OF JUPITER OR SATURN AND CALL IT A 'PLANET', THE DEFINITION OF THIS WORD IS MUCH MORE SUBTLE THAN THAT – AND IT HAS CHANGED OVER TIME

WORDS: ROBERT ROY BRITT

Ever since the discovery of Pluto in 1930, kids grew up learning about the nine planets of our solar system. That all changed starting in the late-1990s, when astronomers began to argue about whether Pluto was a planet. In a highly controversial decision, the International Astronomical Union ultimately decided in 2006 to classify Pluto as a "dwarf planet," reducing the list of "real planets" in our solar system to eight. However, astronomers are now hunting for another planet in our solar system, a true ninth planet, after evidence of its existence was unveiled on 20 Jan 2016. The so-called "Planet Nine," as scientists are calling it, is about ten times the mass of Earth and 5,000 times the mass of Pluto.

The order of the planets, starting nearest the Sun and working outward through the solar system is: Mercury, Venus, Earth, Mars, Jupiter, Saturn, Uranus, Neptune – and Planet Nine. If you insist on including Pluto, then that world would come after Neptune on the list; Pluto is truly way out there, and on a wildly tilted, elliptical orbit (two of the several reasons it got demoted). Interestingly, Pluto used to be the eighth planet. More on that below.

An artist's impression of the theoretical Planet X, whose existence has never been proven

> "CERES WAS ACTUALLY CONSIDERED A PLANET WHEN IT WAS FIRST DISCOVERED IN 1801"

TERRESTRIAL PLANETS

The inner four worlds are called "terrestrial planets," because, like Earth, their surfaces are all rocky. Pluto, too, has a solid surface (and a very frozen one) but has never been grouped with the four terrestrials.

JOVIAN PLANETS

The four large outer worlds – Jupiter, Saturn, Uranus, and Neptune – are known as the "Jovian planets" (meaning "Jupiter-like") because they are all huge compared to the terrestrial planets, and because they are gaseous in nature rather than having rocky surfaces (though some or all of them may have solid cores, astronomers say). Two of the outer planets beyond the orbit of Mars – Jupiter and Saturn – are known as gas giants; the more distant Uranus and Neptune are called ice giants. This is because, while the first two are dominated by gas, the last two have more ice. All four of the outer planets contain mostly hydrogen and helium.

DWARF PLANETS

The IAU definition of a full-fledged planet goes like this: A body that circles the Sun without being some other object's satellite, is large enough to be rounded by its own gravity (but not so big that it begins to undergo nuclear fusion, like a star) and has "cleared its neighborhood" of most other orbiting bodies. Yeah, that's a mouthful.

The problem for Pluto, besides its small size and offbeat orbit, is that it shares its space with lots of other objects in the Kuiper Belt, beyond Neptune. Still, the demotion of Pluto remains controversial.

The IAU planet definition puts other small, round worlds in the dwarf planet category, including the Kuiper Belt objects Eris, Haumea, and Makemake.

Another newly discovered dwarf planet is Ceres, a round object in the Asteroid Belt between Mars and Jupiter. Ceres was actually considered a planet when it was first discovered in 1801, but then later deemed to be an asteroid. Some astronomers like to consider Ceres as a tenth planet – not to be confused with Nibiru or Planet X – but that line of thinking opens up the possibility of there being 13 planets, with more bound to be discovered.

An artist's impression of the solar system

DID YOU KNOW...?

All the planets are named after Roman deities except Uranus, which is named after a Greek god

MEET THE PLANETS

THE EIGHT PRIMARY PLANETS IN OUR SOLAR SYSTEM ALL HAVE THEIR OWN UNIQUE FEATURES, FROM MERCURY OUT TO DISTANT NEPTUNE

MERCURY

The closest planet to the Sun, Mercury is only a bit larger than Earth's Moon. Its day side is scorched by the Sun and can reach 840 degrees Fahrenheit (450 Celsius), but on the night side, temperatures drop to hundreds of degrees below freezing. Mercury has virtually no atmosphere to absorb meteor impacts, so its surface is pockmarked with craters, just like the Moon. Over its four-year mission, NASA's MESSENGER spacecraft has revealed views of the planet that have challenged astronomers' expectations.

STATS

DISCOVERY: KNOWN TO THE ANCIENTS AND VISIBLE TO THE NAKED EYE

NAMED FOR: MESSENGER OF THE ROMAN GODS

DIAMETER: 3,031 MILES (4,878 KM)

ORBIT: 88 EARTH DAYS

LENGTH OF DAY: 58.6 EARTH DAYS

VENUS

The second planet from the Sun, Venus is terribly hot, even hotter than Mercury. The atmosphere is toxic. The pressure at the surface would crush and kill you. Scientists describe Venus' situation as a runaway greenhouse effect. Its size and structure are similar to Earth, Venus' thick, toxic atmosphere traps heat in, creating a climactic hellscape. Oddly, Venus spins slowly in the opposite direction of most planets.

STATS

DISCOVERY: KNOWN TO THE ANCIENTS AND VISIBLE TO THE NAKED EYE

NAMED ROMAN GODDESS OF LOVE AND BEAUTY

DIAMETER: 7,521 MILES (12,104 KM)

ORBIT: 225 EARTH DAYS

LENGTH OF DAY: 241 EARTH DAYS

EARTH

The third planet from the Sun, Earth is a waterworld, with two-thirds of the planet covered by ocean. It's the only world known to harbor life. Earth's atmosphere is rich in life-sustaining nitrogen and oxygen. Earth's surface rotates about its axis at 1,532 feet per second (467 meters per second) – slightly more than 1,000 mph (1,600 kph) – at the equator. The planet zips around the Sun at more than 18 miles per second (29 km per second).

STATS

DIAMETER: 7,926 MILES (12,760 KM)

ORBIT: 365.24 DAYS

LENGTH OF DAY: 23 HOURS, 56 MINUTES

MARS

The fourth planet from the Sun, Mars is a cold, dusty place. The dust, an iron oxide, gives the planet its reddish cast. Mars is rocky, has mountains and valleys, and storm systems ranging that can grow to planet-engulfing dust storms. It snows there, and it harbors water ice. Scientists think it was once wet and warm. Mars' atmosphere is too thin for liquid water to exist on the surface for any length of time today, but scientists believe ancient Mars would have had the conditions to support life, and hope signs of past life may exist on the Red Planet.

STATS

DISCOVERY: KNOWN TO THE ANCIENTS AND VISIBLE TO THE NAKED EYE

NAMED FOR: ROMAN GOD OF WAR

DIAMETER: 74,217 (6,787 KM)

ORBIT: 686 EARTH DAYS

LENGTH OF DAY: 24 HOURS, 37 MINUTES

JUPITER

The fifth planet from the Sun, Jupiter is the largest planet in our solar system. It's a gaseous world, mostly hydrogen and helium. Its swirling clouds are colorful due to different types of trace gases. The Great Red Spot is a giant storm that has raged for centuries. Jupiter has a strong magnetic field and dozens of moons.

STATS

DISCOVERY: KNOWN TO THE ANCIENTS AND VISIBLE TO THE NAKED EYE

NAMED FOR: RULER OF THE ROMAN GODS

DIAMETER: 86,881 MILES (139,822 KM)

ORBIT: 11.9 EARTH YEARS

LENGTH OF DAY: 9.8 EARTH HOURS

URANUS

The seventh planet from the Sun, Uranus is an oddball. It's the only giant planet whose equator is nearly at right angles to its orbit – it basically orbits on its side. Astronomers think the planet collided with some other planet-size object long ago, causing the tilt. The tilt causes extreme seasons that last 20-plus years, and the Sun beats down on one pole or the other for 84 Earth years. Methane in the atmosphere gives Uranus its blue-green tint.

STATS

DISCOVERY: 1781 BY WILLIAM HERSCHEL

NAMED FOR: ANCIENT GREEK KING OF THE GODS

DIAMETER: 31,763 MILES (51,120 KM)

ORBIT: 84 EARTH YEARS

LENGTH OF DAY: 18 EARTH HOURS

SATURN

The sixth planet from the Sun is known most for its rings. When Galileo Galilei first studied Saturn in the early-1600s, he thought it was an object with three parts. Not knowing he was seeing a planet with rings, the stumped astronomer entered a small drawing – a symbol with one large circle and two smaller ones – in his notebook, as a noun in a sentence describing his discovery. More than 40 years later, Christiaan Huygens proposed that they were rings. The rings are made of ice and rock. Scientists are not yet sure how they formed. This gaseous planet consists mostly of hydrogen and helium. It has numerous moons.

STATS

DISCOVERY: KNOWN TO THE ANCIENTS AND VISIBLE TO THE NAKED EYE

NAMED FOR: ROMAN GOD OF AGRICULTURE

DIAMETER: 74,900 MILES (120,500 KM)

ORBIT: 29.5 EARTH YEARS

LENGTH OF DAY: ABOUT 10.5 EARTH HOURS

NEPTUNE

The eighth planet from the Sun, Neptune is known for extreme, sometimes supersonic winds. The freezing cold planet is more than 30 times farther from the Sun than Earth. Neptune was the first planet to be predicted to exist by using math, before it was detected. Irregularities in the orbit of Uranus led French astronomer Alexis Bouvard to suggest another body might be exerting a gravitational tug. German astronomer Johann Galle used calculations to help find Neptune in a telescope.

STATS

DISCOVERY: 1846

NAMED FOR: ROMAN GOD OF WATER

DIAMETER: 30,775 (49,530 KM)

ORBIT: 165 EARTH DAYS

LENGTH OF DAY: 19 EARTH DAYS

PLUTO (DWARF PLANET)

Pluto is unlike the 'true' planets in many respects. Smaller than Earth's Moon, its orbit carries it inside the orbit of Neptune and then way out beyond it. From 1979 until early-1999, Pluto had actually been the eighth planet from the sun. On 11 Feb 1999, it crossed Neptune's path and once again became the solar system's most distant planet – until it was demoted to dwarf planet status. Pluto will stay beyond Neptune for 228 years. Its orbit is tilted to the main plane of the solar system by 17.1 degrees. It's a cold, rocky world with only a very ephemeral atmosphere. NASA's New Horizons mission performed history's first flyby of Pluto on 14 July 2015.

PLANET NINE

Planet Nine orbits the Sun at an orbit 20 times farther out than that of Neptune. Scientists have never seen Planet Nine directly, but its existence was inferred by its gravitational effects on other objects in the distant Kuiper Belt, home to icy objects left over from the birth of the Sun and its planets. Scientists Mike Brown and Konstantin Batygin at the California Institute of Technology described the evidence for Planet Nine in a study published in the *Astronomical Journal*. The research is based on mathematical models and computer simulations using observations of six other smaller Kuiper Belt Objects with orbits that aligned in a similar way.

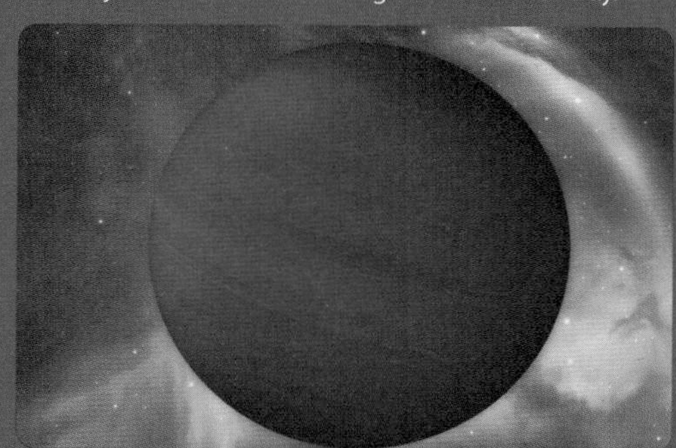

"THE DWARF PLANET PLUTO IS UNLIKE THE 'TRUE' PLANETS IN MANY RESPECTS"

LUNAR CRATERS

WHAT MOON CRATERS CAN TELL US ABOUT EARTH AND OUR SOLAR SYSTEM

WORDS: MEGHAN BARTELS

Asteroid impacts have a bad reputation here on Earth – it's the dinosaurs' signature public relations victory – but it's the moon that really bears the scars of living in our messy neighborhood. That's because Earth has an arsenal of forces that slowly wear away the craters left behind by impacts. And that's frustrating for scientists who want to better understand the debris hurtling around our solar system. So a 2019 study uses the pockmarked lunar surface to trace the history of things smashing into both our Moon and Earth, finding signs that our neighborhood got a lot messier about 290 million years ago.

"It's a cool study that talks about our dynamic solar system and it's good that it's out there," Nicolle Zellner, a physicist at Albion College in Michigan (not involved in the research), told Space.com. "It'll get people thinking and testing it, so that's exciting."

Earth and the Moon are close enough on the solar system scale that stray asteroids should crash into each at about the same frequency. Earth may attract a few extra with its stronger gravity, and Earth likely suffers more hits because of its larger surface area – but in terms of impact per square mile, they should be clocking in about the same.

Scientists have identified only about 180 impact craters here on Earth, as opposed to hundreds of thousands of lunar impact craters. Earth wipes them away with winds and rainfall, oceans and plate tectonics. "The Moon is perfect for studying craters," Sara Mazrouei, a planetary scientist who led the new research during her doctoral studies at the University of Toronto, told Space.com. "Everything stays there."

But in order to trace the history of impacts, scientists needed to not just identify craters, but also estimate their ages. And that's much harder on the Moon than on Earth, since geologists can't currently sample lunar craters directly.

So the team behind the new research settled on what may be a surprising measurement: how well nearby rocks retain heat during the long, cold lunar night. That might seem like an awfully random measurement. But when a large impactor strikes the Moon, it scoops out a crater and litters the surrounding landscape with boulders sourced from that material. Over time, those rocks are struck by smaller impactors that break them into smaller and smaller rocks, which eventually become dusty regolith, so the team argued that older craters would be surrounded by finer rocks and younger craters by larger ones.

A comparison of the near (right) and far (left) side of the Moon

Then, when that landscape transitions from a 14-day lunar day to a 14-day lunar night, it changes temperature at different rates. "The idea is that big rocks can hold heat throughout the night, whereas that regolith or sand loses heat," Mazrouei said. "As craters get older, they become less rocky." In turn, they cool off faster.

So Mazrouei and her colleagues looked at thermal imaging data from an instrument called the Diviner on board NASA's Lunar Reconnaissance Orbiter, which has been circling the Moon since 2009. The team identified 111 individual craters that they knew were less than 1 billion years old, analyzed their heat signatures and, using a model of how quickly lunar boulders disintegrate, estimated their age.

The result showed an intriguing pattern: a spike in impact rates about 290 million years ago, when cratering rates appear to have more than doubled. That would suggest something significant changed in our solar system around then – perhaps, the team proposes, a large space rock in the Asteroid Belt breaking up and wandering closer to Earth and the Moon. And comparing the craters we do know about here on Earth to their results, the team sees similar patterns, suggesting scientists have found a pretty representative, if small, collection of craters.

Not everyone is convinced, however. "The results are intriguing, but I think that the actual support for these conclusions is pretty weak," Jay Melosh, a planetary scientist at Purdue University who wasn't involved in the new research, told Space.com. In particular, he's not sold on the boulder- disintegration model they used – he thinks it doesn't properly account for how that process speeds up as rocks get smaller. And he doesn't see enough Earth craters to support solid statistical analyses; he worries that they're working from too small a sample size. "That doesn't mean it's wrong, but it also doesn't mean that it's right – we just really don't know," Melosh said. "This is a noble attempt to go just a little bit farther than the data support."

Zellner understands how difficult studying lunar craters can be: she's worked with the droplets of glass created by impacts and carried back to Earth in samples gathered by the Apollo astronauts. But dating that glass is still a challenge even with today's lab technology, and the samples all come from a small patch of the Moon's surface. Orbiter data puts scientists at more of a distance, but covers the entire lunar surface – neither method is perfect.

"We're doing the best we can with what we have now," Zellner said. "This is science, right? We put ideas out there, and then we find ways to test those ideas, and the idea either stands the test of time or it doesn't." And all three scientists offered compelling reasons why doing the work to figure out the Moon's impact history is worthwhile. First, of course, there's the self-interested approach: Earth craters can come with some unpleasant side effects. "Everybody is interested in the cratering rate on Earth because we don't want to end up like the dinosaurs," Melosh said. The catastrophic aftermath of the impact wiped out a staggering three-quarters of all species alive at the time, although this mass extinction left plenty of room for our own mammalian ancestors to thrive and evolve. "We should thank our lucky meteorite, but it was pretty bad for everybody else on the planet." Learn enough about impacts, the theory goes, and we may be able to save our own skins the next time around. For Zellner, there's a more exotic appeal as well: Learning more about our own solar system could help scientists understand not just our own neighborhood, but also the processes that have shaped the many alien solar systems that scientists keep discovering.

Mazrouei sees the work as an example of how different solar system bodies can shed light on each other. One of her co-authors is already looking forward to how the BepiColombo mission to Mercury, armed with an instrument much like that at the moon now, will be able to add another dimension to cratering studies.

Earth is great to live on, but scientists can't piece together its past from home. It takes studying the Moon and its pristinely cratered surface to understand what our planet has been through, Mazrouei said. "We get to detangle a lot of Earth's history as well."

"LEARN ENOUGH ABOUT IMPACTS, AND WE MAY BE ABLE TO SAVE OUR OWN SKINS THE NEXT TIME AROUND"

Credit: NASA MSFC

MARS

WHAT WE KNOW ABOUT THE RED PLANET

WORDS: CHARLES Q CHOI

Mars is the fourth planet from the sun. Befitting the Red Planet's bloody color, the Romans named it after their god of war. In truth, the Romans copied the ancient Greeks, who also named the planet after their god of war, Ares. Other civilizations also typically gave the planet names based on its color – for example, the Egyptians named it "Her Desher," meaning "the red one," while ancient Chinese astronomers dubbed it "the fire star."

The bright rust color Mars is known for is due to iron-rich minerals in its regolith – the loose dust and rock covering its surface. The soil of Earth is a kind of regolith, too, albeit one loaded with organic content. According to NASA, the iron minerals oxidize, or rust, causing the soil to look red.

The cold, thin atmosphere means liquid water likely cannot exist on the Martian surface for any length of time. Features called recurring slope lineae may have spurts of briny water flowing on the surface, but this evidence is disputed; some scientists argue the hydrogen spotted from orbit in this region may instead indicate briny salts. This means that although this desert planet is just half the diameter of Earth, it has the same amount of dry land.

The Red Planet is home to both the highest mountain and the deepest, longest valley in the solar system. Olympus Mons is roughly 17 miles (27 kilometers) high, about three times as tall as Mount Everest, while the Valles Marineris system of valleys – named after the Mariner 9 probe that discovered it in 1971 – reaches as deep as 6 miles (10 km) and runs east-west for roughly 2,500 miles (4,000 km), about one-fifth of the distance around Mars and close to the width of Australia.

Scientists think the Valles Marineris formed mostly by rifting of the crust as it got stretched. Individual canyons within the system are as much as 60 miles (100 km) wide. The canyons merge in the central part of the Valles Marineris in a region as much as 370 miles (600 km) wide. Large channels emerging from the ends of some canyons and layered sediments within suggest the canyons might once have been filled with liquid water.

DID YOU KNOW...?

A Martian day is called a "sol", and lasts for 24 hours and 37 minutes

Mars also has the largest volcanoes in the solar system, Olympus Mons being one of them. The massive volcano, which is about 370 miles (600 km) in diameter, is wide enough to cover the state of New Mexico. Olympus Mons is a shield volcano, with slopes that rise gradually like those of Hawaiian volcanoes, and was created by eruptions of lavas that flowed for long distances before solidifying. Mars also has many other kinds of volcanic landforms, from small, steep-sided cones to enormous plains coated in hardened lava. Some minor eruptions might still occur on the planet.

Channels, valleys and gullies are found all over Mars, and suggest that liquid water might have flowed across the planet's surface in recent times. Some channels can be 60 miles (100 km) wide and 1,200 miles (2,000 km) long. Water may still lie in cracks and pores in underground rock. A study by

INTERNAL STRUCTURE

Scientists think that on average, the Martian core is between 1,800 and 2,400 miles in diameter (3,000 and 4,000 km), its mantle is about 900 to 1,200 miles (5,400 to 7,200 km) wide and its crust is about 30 miles (50 km) thick.

COMPOSITION

95.32% CARBON DIOXIDE
2.7% NITROGEN
1.6% ARGON
0.13% OXYGEN
0.08% CARBON MONOXIDE

With minor amounts of water, nitrogen oxide, neon, hydrogen-deuterium-oxygen, krypton and xenon.

MARS' ORBIT

AVERAGE DISTANCE FROM THE SUN:
141,633,260 MILES (227,936,640 KM)
1.524 TIMES THAT OF EARTH

PERIHELION (CLOSEST):
128,400,000 MILES (206,600,000 KM)
1.404 TIMES THAT OF EARTH

APHELION (FARTHEST):
154,900,000 MILES (249,200,000 KM)
1.638 TIMES THAT OF EARTH

Images taken by the Mars Reconnaissance Orbiter show the Red Planet before (left) and after (right) a dust storm engulfed the planet in 2018

scientists in 2018 suggested that salty water below the Martian surface could hold a considerable amount of oxygen, which would support microbial life. However, the amount of oxygen depends on temperature and pressure, as well as temperature changes on Mars from time to time as the tilt of its rotation axis shifts.

Many regions of Mars are flat, low-lying plains. The lowest of the northern plains are among the flattest, smoothest places in the solar system, potentially created by water that once flowed across the Martian surface. The northern hemisphere mostly lies at a lower elevation than the southern hemisphere, suggesting the crust may be thinner in the north than in the south. This difference between the north and south might be due to a very large impact shortly after the birth of Mars.

The number of craters on Mars varies dramatically from place to place, depending on how old the surface is. Much of the surface of the southern hemisphere is extremely old, and so has many craters – including the planet's largest, 1,400-mile-wide (2,300 km) Hellas Planitia – while that of northern hemisphere is younger and so has fewer craters. Some volcanoes also have a few craters, which suggests they erupted recently, with the resulting lava covering up any old craters. Some craters have unusual-looking deposits of debris around them resembling solidified mudflows, potentially indicating that the impactor hit underground water or ice.

In 2018, the European Space Agency's Mars Express spacecraft detected what could be a slurry of water and grains underneath icy Planum Australe. (Some reports describe it as a "lake," but it's unclear how much regolith is inside the water.) This body of water is said to be about 12.4 miles (20 km) across. Its underground location is reminiscent of similar underground lakes in Antarctica, which have been found to host microbes. Late in the year, Mars Express also spied a huge, icy zone in the Red Planet's Korolev Crater.

POLAR CAPS

Vast deposits of what appear to be finely layered stacks of water ice and dust extend from the poles to latitudes of about 80 degrees in both hemispheres – most likely deposited by the atmosphere over long spans of time. On top of much of these layered deposits in both hemispheres are caps of water ice that remain frozen year-round.

Additional seasonal caps of frost appear in winter. These are made of solid carbon dioxide, also known as "dry ice," which has condensed from carbon dioxide gas in the atmosphere. In the deepest part of the winter, this frost can extend from the poles to latitudes as low as 45 degrees, or halfway to the equator. The dry ice layer appears to have a fluffy texture, like freshly fallen snow, according to a report in the Journal of Geophysical Research-Planets.

Credit: NASA/JPL-Caltech/MSSS

MAGNETIC FIELD

Mars currently has no global magnetic field, but there are regions of its crust that can be at least ten times more strongly magnetized than anything that has been measured on Earth, which suggests those regions are remnants of an ancient global magnetic field.

Credit: NASA/GSFC

CHEMICAL COMPOSITION

Mars likely has a solid core composed of iron, nickel and sulfur. The mantle of Mars is probably similar to Earth's in that it is composed mostly of peridotite, which is made up primarily of silicon, oxygen, iron and magnesium. The crust is probably largely made of the volcanic rock basalt, which is also common in the crusts of the Earth and the moon, although some crustal rocks, especially in the northern hemisphere, may be a form of andesite, a volcanic rock that contains more silica than basalt does.

Credit: NASA/JPL-Caltech/GSFC/Univ. of Arizona

"SALTY WATER BELOW THE MARTIAN SURFACE COULD HOLD OXYGEN"

Credit: NASA/JPL-Caltech/MSSS

NASA have sent many orbiters, probes and rovers to study the Martian surface

CLIMATE

Mars is much colder than Earth, in large part due to its greater distance from the sun. The average temperature is about minus 80 degrees Fahrenheit (minus 60 degrees Celsius), although it can vary from minus 195 F (minus 125 C) near the poles during the winter to as much as 70 F (20 C) at midday around the equator.

The carbon-dioxide-rich atmosphere of Mars is also about 100 times less dense than Earth's on average, but it is nevertheless thick enough to support weather, clouds and winds. The density of the atmosphere varies seasonally, as winter forces carbon dioxide to freeze out of the Martian air. In the ancient past, the atmosphere was likely thicker and able to support water flowing on its surface. Over time, lighter molecules in the Martian atmosphere escaped under pressure from the solar wind, which affected the atmosphere because Mars does not have a global magnetic field. This process is being studied today by NASA's MAVEN (Mars Atmosphere and Volatile Evolution) mission.

NASA's Mars Reconnaissance Orbiter found the first definitive detections of carbon-dioxide snow clouds, making Mars the only body in the solar system known to host such unusual winter weather. The Red Planet also causes water-ice snow to fall from the clouds.

The dust storms on Mars are the largest in the solar system, able to blanket the entire Red Planet and last months. One theory as to why dust storms can grow so big on Mars is because the airborne dust particles absorb sunlight, warming the Martian atmosphere in their vicinity. Warm pockets of air then flow toward colder regions, generating winds. Strong winds lift more dust off the ground, which, in turn, heats the atmosphere, raising more wind and kicking up more dust.

"DUST STORMS ON MARS ARE THE LARGEST IN THE SOLAR SYSTEM, ABLE TO BLANKET THE ENTIRE RED PLANET AND LAST MONTHS"

ORBITAL CHARACTERISTICS

The axis of Mars, like Earth's, is tilted with relation to the sun. This means that like Earth, the amount of sunlight falling on certain parts of the Red Planet can vary widely during the year, giving Mars seasons.

However, the seasons that Mars experiences are more extreme than Earth's because the Red Planet's elliptical, oval-shaped orbit around the sun is more elongated than that of any of the other

THE MOONS OF MARS

WHAT DO WE KNOW ABOUT THE RED PLANET'S NATURAL SATELLITES?

The two moons of Mars, Phobos and Deimos, were discovered by American astronomer Asaph Hall over the course of a week in 1877. Hall had almost given up his search for a moon of Mars, but his wife, Angelina, urged him on. He discovered Deimos the next night, and Phobos six days after that. He named the moons after the sons of the Greek war god Ares – Phobos means "fear," while Deimos means "rout."

Both Phobos and Deimos are apparently made of carbon-rich rock mixed with ice and are covered in dust and loose rocks. They are tiny next to Earth's moon, and are irregularly shaped, since they lack enough gravity to pull themselves into a more circular form. The widest Phobos gets is about 17 miles (27 km), and the widest Deimos gets is roughly 9 miles (15 km).

Both moons are pockmarked with craters from meteor impacts. The surface of Phobos also possesses an intricate pattern of grooves, which may be cracks that formed after the impact created the moon's largest crater – a hole about 6 miles (10 km) wide, or nearly half the width of Phobos. They always show the same face to Mars, just as our moon does to Earth.

It remains uncertain how Phobos and Deimos formed. They may have been asteroids captured by Mars' gravitational pull, or formed in orbit around Mars the same time as the planet. Ultraviolet light reflected from Phobos provides strong evidence that the moon is a captured asteroid, according to astronomers at the University of Padova in Italy.

Phobos is gradually spiraling toward Mars, drawing about 6 feet (1.8 m) closer to the Red Planet each century. Within 50 million years, Phobos will either smash into Mars or break up and form a ring of debris around the planet.

Many scientists believe that Mars's moons Phobos (center) and Deimos (right) are captured asteroids

Credit: NASA/JPL-Caltech/GSFC/Univ. of Arizona

major planets. When Mars is closest to the sun, its southern hemisphere is tilted toward the sun, giving it a short, very hot summer, while the northern hemisphere experiences a short, cold winter. When Mars is farthest from the sun, the northern hemisphere is tilted toward the sun, giving it a long, mild summer, while the southern hemisphere experiences a long, cold winter.

The tilt of the Red Planet's axis swings wildly over time because it's not stabilized by a large moon, like Earth is. This led to different climates on the Martian surface throughout its history. A 2017 study suggests that the changing tilt also influenced the release of methane into Mars' atmosphere, causing temporary warming periods that allowed water to flow.

RESEARCH & EXPLORATION

The first person to observe Mars with a telescope was Galileo Galilei. In the century following, astronomers discovered the planet's polar ice caps. In the 19th and 20th centuries, researchers believed they saw a network of long, straight canals on Mars, that hinted at possible civilization, although later these proved to be mistaken interpretations of dark regions they saw.

A number of martian rocks have fallen to Earth over the years, providing scientists a rare opportunity to study Martian rocks without having to leave our planet. One of the most controversial finds was Allan Hills 84001 – a Martian meteorite that in 1996, was said to contain shapes reminiscent of small fossils. The find garnered a lot of media attention at the time, but subsequent studies dismissed the idea. In 2018, a separate meteorite study found that organic molecules – the building blocks of life, although not necessarily life itself – could have formed on Mars through battery-like chemical reactions.

Robotic spacecraft began observing Mars in the 1960s, with the United States launching Mariner 4 in 1964 and Mariners 6 and 7 in 1969. These missions revealed the Red Planet to be a barren world, without any signs of the life or the alien civilizations people had imagined could be hiding there. In 1971, Mariner 9 orbited Mars, mapping about 80 percent of the planet and revealing its volcanoes and canyons.

A shot of the desolate Martian surface, taken by the Spirit rover near its landing site

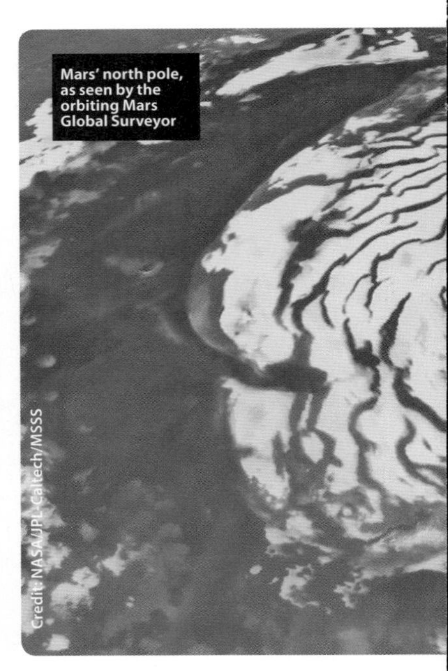

Mars' north pole, as seen by the orbiting Mars Global Surveyor

"NASA PLANS TO LAUNCH A SUCCESSOR TO CURIOSITY, CALLED MARS 2020, WHICH WILL SEARCH FOR ANCIENT SIGNS OF LIFE ON THE RED PLANET"

The Soviet Union also launched numerous spacecraft in the 1960s and early 1970s, but most of those missions failed. Mars 2 (1971) and Mars 3 (1971) operated successfully, but were unable to map the surface due to dust storms. NASA's Viking 1 lander touched down on the surface of Mars in 1976, the first successful landing on the Red Planet. The lander took the first close-up pictures of the Martian surface but found no strong evidence for life.

The next two craft to successfully reach Mars were the Mars Pathfinder, a lander, and Mars Global Surveyor, an orbiter, both launched in 1996. A small robot onboard Pathfinder named Sojourner – the first wheeled rover to explore the surface of another planet – ventured across the planet's surface analyzing rocks.

In 2001, NASA launched the Mars Odyssey orbiter which reached the Red Planet in October that year. Odyssey discovered vast amounts of water ice beneath the Martian surface, mostly in the upper 3 feet (1 meter). It remains uncertain whether more water lies underneath, since the probe cannot see water any deeper.

In 2003, Mars passed closer to Earth than anytime in that past 60,000 years. That same year, NASA launched two rovers, nicknamed Spirit and Opportunity, which explored different regions of the Martian surface. Both rovers found signs that water once flowed on the planet's surface.

In 2008, NASA sent another mission, Phoenix, to land in the northern plains of Mars and search for water which it succeeded in doing.

In 2011, NASA's Mars Science Laboratory mission sent the Mars Curiosity rover, to investigate Martian rocks and determine the geologic processes that created them. Among the mission's findings was the first meteorite on the surface of the Red Planet. The rover has found complex organic molecules on the surface, as well as seasonal fluctuations in methane concentrations in the atmosphere.

NASA has two other orbiters working around the planet, Mars Reconnaissance Orbiter and MAVEN (Mars Atmosphere and Volatile Evolution). The European Space Agency (ESA) also has two spacecraft orbiting the planet: Mars Express and the Trace Gas Orbiter.

In September 2014, India's Mars Orbiter Mission also reached the Red Planet, making it the fourth nation to successfully enter orbit around Mars.

In May 2018, NASA sent a stationary lander called Mars InSight to the surface. Since landing in November that year, InSight has examined geologic activity by burrowing a probe underground.

NASA plans to launch a successor rover mission to Curiosity, called Mars 2020. This mission will search for ancient signs of life and, depending on how promising its samples look, it may "cache" the results in safe spots on the Red Planet for a future rover to pick up.

ESA is working on its own ExoMars rover that should also launch in 2020, and will include a drill to go deep into the Red Planet, collecting soil samples from about 2 meters (6.5 feet) deep.

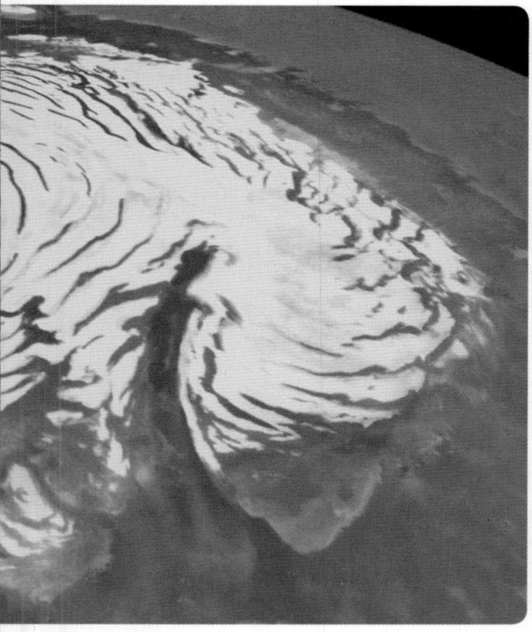

HUMAN MISSIONS TO MARS

HOW WILL THE FIRST ASTRONAUTS MAKE IT TO THE RED PLANET?

Robots aren't the only ones getting a ticket to Mars. A workshop group of scientists from government agencies, academia and industry have determined that a NASA-led manned mission to Mars should be possible by the 2030s. However, in late 2017, the Trump administration directed NASA to send people back to the moon before going to Mars. NASA is now more focused on a concept called the Lunar Orbital Platform-Gateway that would be a moon-based space station and headquarters for further space exploration.

Robotic missions to the Red Planet have seen much success in the past few decades, but it remains a considerable challenge to get people to Mars. With current rocket technology, it would take several months for people to travel to Mars, and that means they would live for several months in microgravity, which has devastating effects on the human body. Performing activities in the moderate gravity on Mars could prove extremely difficult after many months in microgravity. Research on the effects of microgravity continues on the International Space Station.

NASA isn't the only one with Martian astronaut hopefuls. Elon Musk, the founder of SpaceX, has outlined multiple concepts to bring people to Mars. In November 2018, Musk rebranded SpaceX's future "Big Falcon Rocket" to "Starship". Other nations, including China and Russia, have also announced their goals for sending humans to Mars.

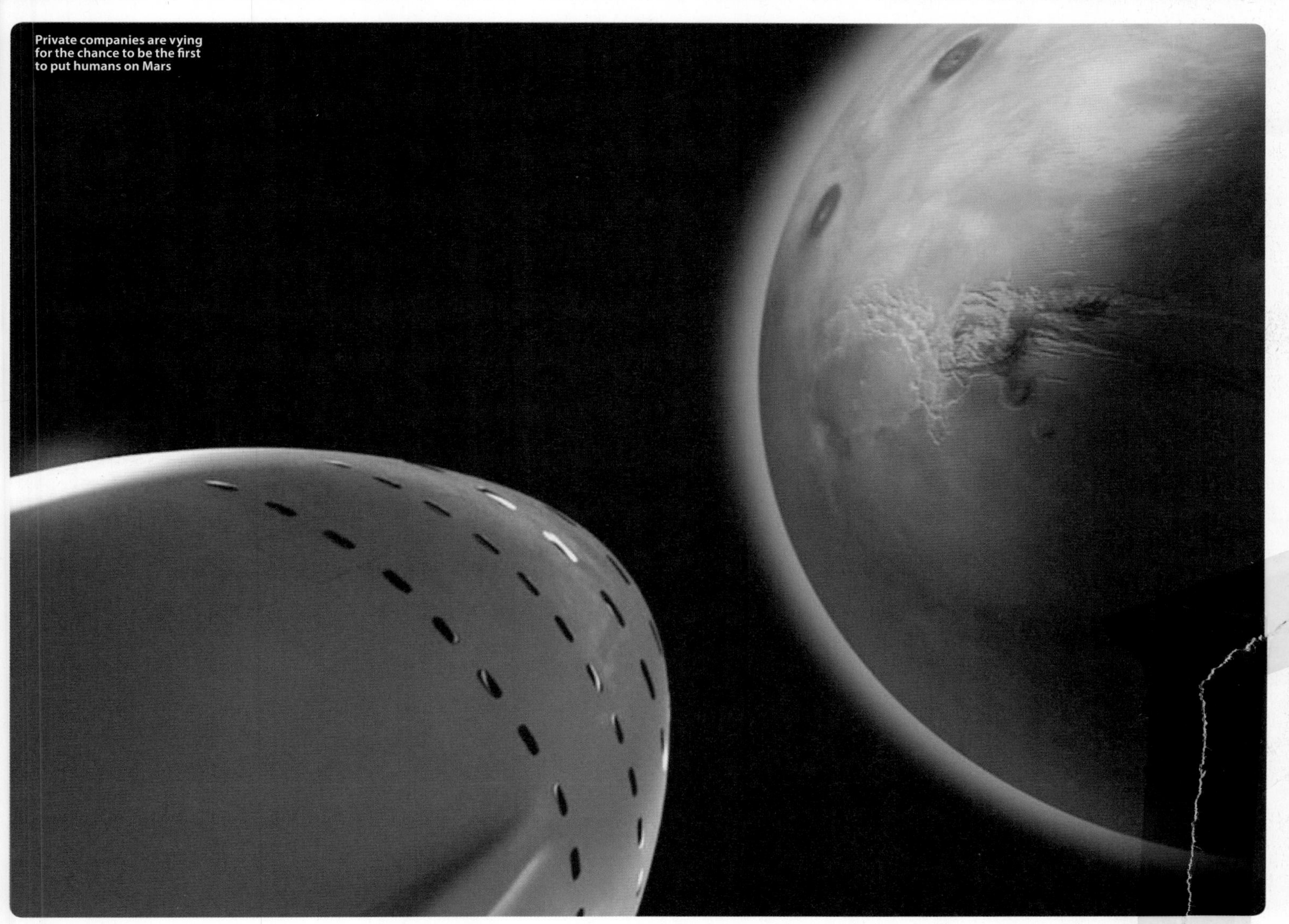

Private companies are vying for the chance to be the first to put humans on Mars

WHAT CAN METEORITES REVEAL?

METEORITE STUDIES OFFER SOME SURPRISES ABOUT EARTH'S FORMATION

WORDS: ELIZABETH HOWELL

Studies in 2017 gave scientists a better understanding of three concepts regarding the early Earth: what kind of raw materials coalesced to form Earth long ago, when water arrived on our planet, and why Earth and its moon have such similar compositions.

Two studies published in Jan. 25 2017 in the journal Nature suggest that Earth's main building blocks were rocks similar to meteorites known as enstatite chondrites, and that the planet got most of its water gradually during the formation process, rather than in one big burst toward the end.

"The results presented in these papers lead to the troubling conclusion that the meteorites in our collection are not particularly good examples of Earth's building blocks," Richard Carlson, a geochemist at the Carnegie Institution for Science in Washington, D.C., wrote in a commentary article that accompanied the two studies in Nature. (Carlson was not involved in either study.)

WHEN DID WATER COME TO EARTH?

Scientists have known since the 1970s that Earth rocks' oxygen-isotope abundance differs from that of most meteorites, except for enstatite chondrites, Carlson wrote. Isotopes are variants of the same element that have different numbers of neutrons in their atomic nuclei.

But Earth rocks and enstatite meteorites differ in elemental composition, so most researchers have used Earth-formation models based on different meteorites called carbonaceous chondrites, which are rich in volatiles (compounds with low boiling points, such as water), he added.

By tracking different isotopic abundances in Earth rocks and meteorites, both studies concluded that carbonaceous-chondrite-like building blocks were not common late in the Earth's formation history.

Specifically, one of the studies – by University of Chicago geochemist Nicolas Dauphas – suggested that several different meteorite types were likely responsible for the first 60 percent of Earth's growth, while almost all of the remaining 40 percent came from enstatite chondrites.

The second study, performed by Mario Fischer-Gödde and Thorsten Kleine of the University of Münster in Germany, bolstered that conclusion, showing that enstatite-like rocks probably dominated Earth's late accretion history.

These findings indicate that water came to Earth throughout the planet's formation history, and not just in a flurry of carbonaceous chondrites and/or comets near the end, as some researchers have proposed, Carlson wrote.

"If the last 0.5 percent of material accreted by Earth had been composed of a particular type of volatile-rich carbonaceous chondrite, known as a CI chondrite, an amount of water equivalent in mass to Earth's oceans would have been added to the planet," Carlson explained. "Fischer-Gödde and Kleine's measurements instead show that this late-accreted material consisted of relatively 'dry' enstatite chondrites."

However, this conclusion still leaves a question: Why doesn't Earth's surface composition match that of enstatite chondrites? Carlson suggested two possible explanations: Earth's deep interior is very different from the surface (which is unlikely for a variety of reasons, he wrote), or, as Dauphas suggested in his paper, the enstatite chondrites were altered on Earth's surface as the planet evolved.

WHY IS THE MOON SO SIMILAR TO EARTH?

Dauphas' study also sheds light on how the moon formed. Most astronomers think that, long ago, one or more Mars-size bodies smashed into the proto-Earth, blasting out material that eventually coalesced to form the moon.

Models suggest that such giant impacts likely would have created a moon whose composition was different from Earth's, because proto-Earth and the impactor(s) likely would have been made of different stuff. But measurements of Earth, the moon and enstatite meteorites "have almost indistinguishable isotopic compositions," Dauphas wrote in his new study.

The apparent problem can be traced to older models that suggest a diversity of meteorites formed Earth, Dauphas added. His research, on the other hand, indicates that the proto-Earth and the moon-forming impactor(s) probably formed in the same "isotopic reservoir," which was dominated by enstatite chondrites.

"Accordingly, the giant impactor that formed the moon probably had an isotopic composition similar to that of the Earth, hence relaxing the constraints on models of lunar formation," Dauphas wrote.

Credit: NASA/JPL-Caltech/MSSS

An iron-nickel meteorite spotted by the Curiosity rover on Mars in 2016

Any space rocks that survive their journey through the atmosphere to hit the ground are known as meteorites

PLANET EARTH

OUR PALE BLUE DOT IS UNIQUE AMONG ANY OF THE OTHER PLANETS WE HAVE DISCOVERED SO FAR

WORDS: CHARLES Q CHOI

Earth, our home, is the third planet from the sun. It's the only planet known to have an atmosphere containing free oxygen, oceans of water on its surface and, of course, life.

Earth is the fifth largest of the planets in the solar system. It's smaller than the four gas giants – Jupiter, Saturn, Uranus and Neptune – but larger than the three other rocky planets, Mercury, Mars and Venus.

Earth has a diameter of roughly 8,000 miles (13,000 kilometers) and is round because gravity pulls matter into a ball. But, it's not perfectly round. Earth is really an "oblate spheroid," because its spin causes it to be squashed at its poles and swollen at the equator.

Water covers roughly 71 percent of Earth's surface, and most of that is in the oceans. About a fifth of Earth's atmosphere consists of oxygen, produced by plants. While scientists have been studying our planet for centuries, much has been learned in recent decades by studying pictures of Earth from space.

DID YOU KNOW...?

Depending on where you are on the globe, you may be spinning through space at over 1,000mph!

EARTH'S ORBIT

While Earth orbits the sun, the planet is simultaneously spinning on an imaginary line called an axis that runs from the North Pole to the South Pole. It takes Earth 23.934 hours to complete a rotation on its axis and 365.26 days to complete an orbit around the sun.

Earth's axis of rotation is tilted in relation to the ecliptic plane, an imaginary surface through the planet's orbit around the sun. This means the Northern and Southern hemispheres will sometimes point toward or away from the sun depending on the time of year. This changes the amount of light the hemispheres receive, resulting in the seasons.

Earth's orbit is not a perfect circle, but rather an oval-shaped ellipse, similar to the orbits of all the other planets. Our planet is a bit closer to the sun in early January and farther away in July, although this variation has a much smaller effect than the heating and cooling caused by the tilt of Earth's axis. Earth happens to lie within the so-called "Goldilocks zone" around the sun, where temperatures are just right to maintain liquid water on our planet's surface.

EARTH'S FORMATION AND EVOLUTION

Scientists think Earth was formed at roughly the same time as the sun and other planets some 4.6 billion years ago, when the solar system coalesced from a giant, rotating cloud of gas and dust known as the solar nebula. As the nebula collapsed because of its gravity, it spun faster and flattened into a disk. Most of the material was pulled toward the center to form the sun.

Other particles within the disk collided and stuck together to form ever-larger bodies, including Earth. Scientists think Earth started off as a waterless mass of rock.

"It was thought that because of these asteroids and comets flying around colliding with Earth, conditions on early Earth may have been hellish," Simone Marchi, a planetary scientist at the Southwest Research Institute in Boulder, Colorado, previously told Space.com. But in recent years, new analyses of minerals trapped within ancient microscopic crystals suggests that there was liquid water already present on Earth during its first 500 million years, Marchi said.

Radioactive materials in the rock and increasing pressure deep within the Earth generated enough heat to melt the planet's interior, causing some

Credit: NASA

DID YOU KNOW...?

Studies have shown that Earth's gravity is "lumpy" – its strength is uneven across the planet's surface

"THE CORE IS RESPONSIBLE FOR EARTH'S MAGNETIC FIELD, WHICH HELPS TO DEFLECT HARMFUL PARTICLES SHOT FROM THE SUN"

chemicals to rise to the surface and form water, while others became the gases of the atmosphere. Recent evidence suggests that Earth's crust and oceans may have formed within about 200 million years after the planet took shape.

INTERNAL STRUCTURE

Earth's core is about 4,400 miles (7,100 km) wide, slightly larger than half the Earth's diameter and about the same size as Mars' diameter. The outermost 1,400 miles (2,250 km) of the core are liquid, while the inner core is solid; it's about four-fifths as big as Earth's moon, at some 1,600 miles (2,600 km) in diameter. The core is responsible for the planet's magnetic field, which helps to deflect harmful charged particles shot from the sun.

Above the core is Earth's mantle, which is about 1,800 miles (2,900 km) thick. The mantle is not completely stiff but can flow slowly. Earth's crust floats on the mantle much as a piece of wood floats on water. The slow motion of rock in the mantle shuffles continents around and causes earthquakes, volcanoes and the formation of mountain ranges.

Above the mantle, Earth has two kinds of crust. The dry land of the continents consists mostly of granite and other light silicate minerals, while the ocean floors are made up mostly of a dark, dense volcanic rock called basalt. Continental crust averages some 25 miles (40 kilometres) thick, although it can be thinner or thicker in some areas. Oceanic crust is usually only about 5 miles (8 kilometres) thick. Water fills in low areas of the basalt crust to form the world's oceans.

Earth gets warmer toward its core. At the bottom of the continental crust, temperatures reach about 1,800 degrees Fahrenheit (1,000 degrees Celsius), increasing about 3 degrees F per mile (1 degree C per km) below the crust. Geologists think the temperature of Earth's outer core is about 6,700 to 7,800 degrees F (3,700 to 4,300 degrees C) and that the inner core may reach 12,600 degrees F (7,000 degrees C) – hotter than the surface of the sun.

MAGNETIC FIELD

Earth's magnetic field is generated by currents flowing in Earth's outer core. The magnetic poles are always on the move, with the magnetic North Pole accelerating its northward motion to 24 miles (40 km) annually since tracking began in the 1830s. It will likely exit North America and reach Siberia in a matter of decades.

Earth's magnetic field is changing in other ways, too. Globally, the magnetic field has weakened 10 percent since the 19th century, according to NASA. These changes are mild compared to what Earth's magnetic field has done in the past. A few times every million years or so, the field completely flips, with the North and the South poles swapping places. The magnetic field can take anywhere from 100 to 3,000 years to complete the flip.

The strength of Earth's magnetic field decreased by about 90 percent when a field reversal occurred in ancient past, according to Andrew Roberts, a professor at the Australian National University. The drop makes the planet more vulnerable to solar storms and radiation, which could significantly damage satellites, disrupting our modern communication and electrical infrastructure.

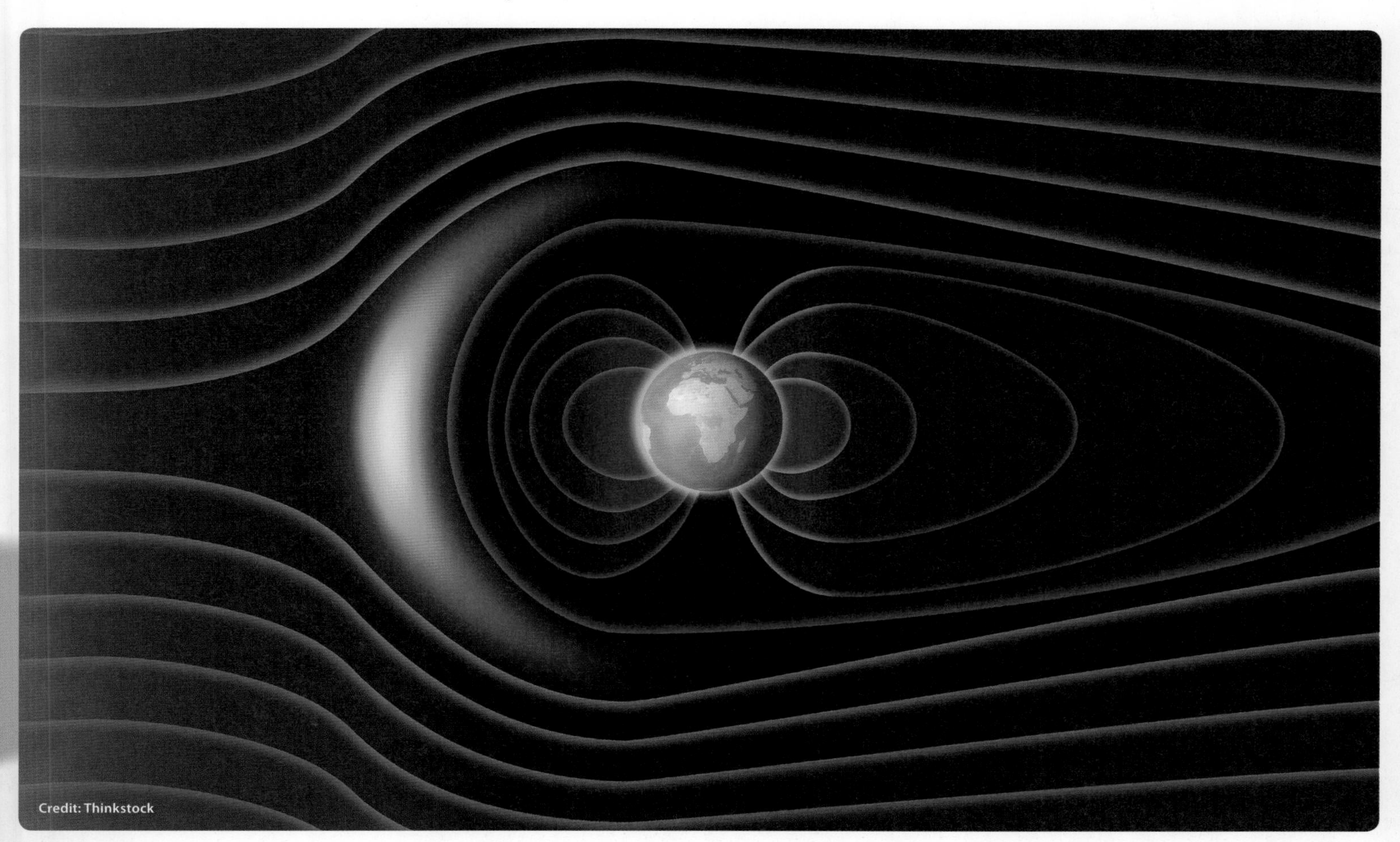
Credit: Thinkstock

"Hopefully, such an event is a long way in the future and we can develop future technologies to avoid huge damage," Roberts said in a statement.

When charged particles from the sun get trapped in Earth's magnetic field, they smash into air molecules above the magnetic poles, causing them to glow. This phenomenon is known as the aurorae, the northern and southern lights.

EARTH'S ATMOSPHERE

Earth's atmosphere is roughly 78 percent nitrogen and 21 percent oxygen, with trace amounts of water, argon, carbon dioxide and other gases. Nowhere else in the solar system is there an atmosphere loaded with free oxygen, which is vital to one of the other unique features of Earth: life (or at least, life as we know it).

Air surrounds Earth and becomes thinner farther from the surface. Roughly 100 miles (160 kilometres) above Earth, the air is so thin that satellites can zip through the atmosphere with little resistance. Still, traces of atmosphere can be found as high as 370 miles (600 kilometres) above the planet's surface.

The lowest layer of the atmosphere is known as the troposphere, which is constantly in motion. This perpetual movement is why we have weather. Sunlight heats the planet's surface, causing warm air to rise into the troposphere. This air expands and cools as air pressure decreases, and because this cool air is denser than its surroundings, it then sinks and gets warmed by the Earth, and the cycle begins again.

Above the troposphere, some 30 miles (48 kilometres) above the Earth's surface, is the stratosphere. The still air of the stratosphere contains the ozone layer, which was created when ultraviolet light caused trios of oxygen atoms to bind together into ozone molecules. Ozone prevents most of the sun's harmful ultraviolet radiation from reaching Earth's surface, where it can damage and mutate life.

Water vapor, carbon dioxide and other gases in the atmosphere trap heat from the sun, warming Earth. Without this so-called "greenhouse effect," Earth would probably be too cold for life to exist, although a runaway greenhouse effect led to the hellish conditions now seen on Venus.

Earth-orbiting satellites show that the upper atmosphere actually expands during the day and contracts at night due to heating and cooling.

CHEMICAL COMPOSITION

Oxygen is the most abundant element in rocks in Earth's crust, composing roughly 47 percent of the weight of all rock. The second most abundant element is silicon, at 27 percent, followed by aluminum, at 8 percent; iron, at 5 percent; calcium, at 4 percent; and sodium, potassium and magnesium, at about 2 percent each.

Earth's core consists mostly of iron and nickel and potentially smaller amounts of lighter elements, such as sulfur and oxygen. The mantle

is made of iron and magnesium-rich silicate rocks. (The combination of silicon and oxygen is known as silica, and minerals that contain silica are known as silicate minerals.)

EARTH'S MOON

Earth's moon is 2,159 miles (3,474 kilometres) wide, about one-fourth of Earth's diameter. Our planet has one moon, while Mercury and Venus have none and all the other planets in our solar system have two or more.

The leading explanation for how Earth's moon formed is that a giant impact knocked the raw ingredients for the moon off the primitive, molten Earth and into orbit. Scientists have suggested that the object that hit the planet had roughly 10 percent the mass of Earth, about the size of Mars.

LIFE ON EARTH

Earth is the only planet in the universe known to possess life. The planet boasts several million species of life, living in habitats ranging from the bottom of the deepest ocean to a few miles into the atmosphere. And scientists think far more species remain to be discovered.

Researchers suspect that other candidates for hosting life in our solar system – such as Saturn's moon Titan or Jupiter's moon Europa – could house primitive living creatures. Scientists have yet to precisely nail down exactly how our primitive ancestors first showed up on Earth. One solution suggests that life first evolved on the nearby planet Mars, once a habitable planet, then traveled to Earth on meteorites hurled from the Red Planet by impacts from other space rocks.

"It's lucky that we ended up here, nevertheless, as certainly Earth has been the better of the two planets for sustaining life," biochemist Steven Benner, of the Westheimer Institute for Science and Technology in Florida, told Space.com. "If our hypothetical Martian ancestors had remained on Mars, there might not have been a story to tell."

EARTH'S ORBITAL STATS

AVERAGE DISTANCE FROM THE SUN:
92,956,050 MILES (149,598,262 KM)

PERIHELION (CLOSEST APPROACH TO THE SUN):
91,402,640 MILES (147,098,291 KM)

APHELION (FARTHEST DISTANCE FROM THE SUN):
94,509,460 MILES (152,098,233 KM)

LENGTH OF SOLAR DAY (SINGLE ROTATION ON ITS AXIS):
23.934 HOURS

LENGTH OF YEAR (SINGLE REVOLUTION AROUND THE SUN):
365.26 DAYS

EQUATORIAL INCLINATION TO ORBIT:
23.4393 DEGREES

Credit: NASA / JSC

THE SUN

ALL ABOUT OUR SOLAR SYSTEM'S STAR, ITS AGE, SIZE AND HISTORY

WORDS: CHARLES Q. CHOI

The sun lies at the heart of the solar system, where it is by far the largest object. It holds 99.8 percent of the solar system's mass and is roughly 109 times the diameter of the Earth – about one million Earths could fit inside the sun.

The visible part of the sun is about 10,000 degrees Fahrenheit (5,500 degrees Celsius), while temperatures in the core reach more than 27 million F (15 million C), driven by nuclear reactions. One would need to explode 100 billion tons of dynamite every second to match the energy produced by the sun, according to NASA.

The sun is one of more than 100 billion stars in the Milky Way. It orbits some 25,000 light-years from the galactic core, completing a revolution once every 250 million years or so. The sun is relatively young, part of a generation of stars known as Population I, which are relatively rich in elements heavier than helium. An older generation of stars is called Population II, and an earlier generation of Population III may have existed, although no members of this generation are known yet.

FORMATION & EVOLUTION

The sun was born about 4.6 billion years ago. Many scientists think the sun and the rest of the solar system formed from a giant, rotating cloud of gas and dust known as the solar nebula. As the nebula collapsed because of its gravity, it spun faster and flattened into a disk. Most of the material was pulled toward the center to form the sun.

The sun has enough nuclear fuel to stay much as it is now for another 5 billion years. After that, it will swell to become a red giant. Eventually, it will shed its outer layers, and the remaining core will collapse to become a white dwarf. Slowly, this will fade, to enter its final phase as a dim, cool theoretical object sometimes known as a black dwarf.

INTERNAL STRUCTURE AND ATMOSPHERE

The sun and its atmosphere are divided into several zones and layers. The solar interior, from the inside out, is made up of the core, radiative zone and the convective zone. The solar atmosphere above that consists of the photosphere, chromosphere, a transition region and the corona. Beyond that is the solar wind, an outflow of gas from the corona.

The core extends from the sun's center to about a quarter of the way to its surface. Although it only makes up roughly 2 percent of the sun's volume, it is almost 15 times the density of lead and holds nearly half of the sun's mass. Next is the radiative zone, which extends from the core to 70 percent of the way to the sun's surface, making up 32 percent of the sun's volume and 48 percent of its mass. Light from the core gets scattered in this zone, so that a single photon often may take about a million years to pass through.

The convection zone reaches up to the sun's surface, and makes up 66 percent of the sun's volume but only a little more than 2 percent of its mass. Roiling "convection cells" of gas dominate this zone. Two main kinds of solar convection cells exist – granulation cells about 600 miles (1,000 kilometers) wide and supergranulation cells about 20,000 miles (30,000 km) in diameter.

The photosphere is the lowest layer of the sun's atmosphere, and emits the light we see. It is about 300 miles (500 km) thick, although most of the light comes from its lowest third. Temperatures in the photosphere range from 11,000 F (6,125 C) at bottom to 7,460 F (4,125 C) at top. Next up is the chromosphere, which is hotter, up to 35,500 F (19,725 C), and is apparently made up entirely of spiky structures known as spicules typically some 600 miles (1,000 km) across and up to 6,000 miles (10,000 km) high.

After that is the transition region a few hundred to a few thousand miles thick, which is heated by the corona above it and sheds most of its light as ultraviolet rays. At the top is the super-hot corona, which is made of structures such as loops and streams of ionized gas. The corona generally ranges

"THE SUN HAS ENOUGH FUEL TO STAY MUCH AS IT IS NOW FOR ANOTHER 5 BILLION YEARS"

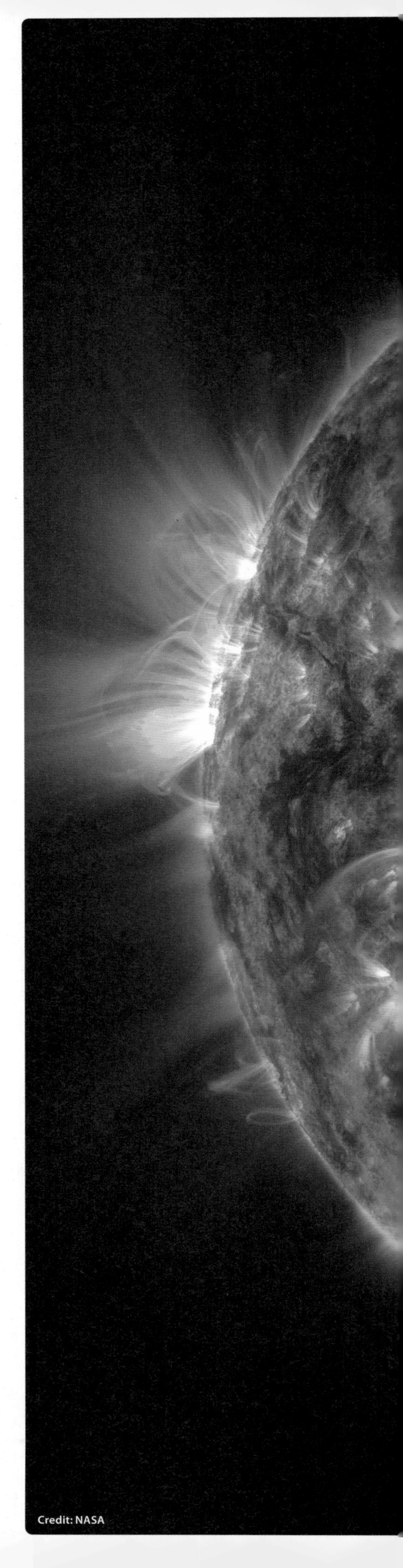
Credit: NASA

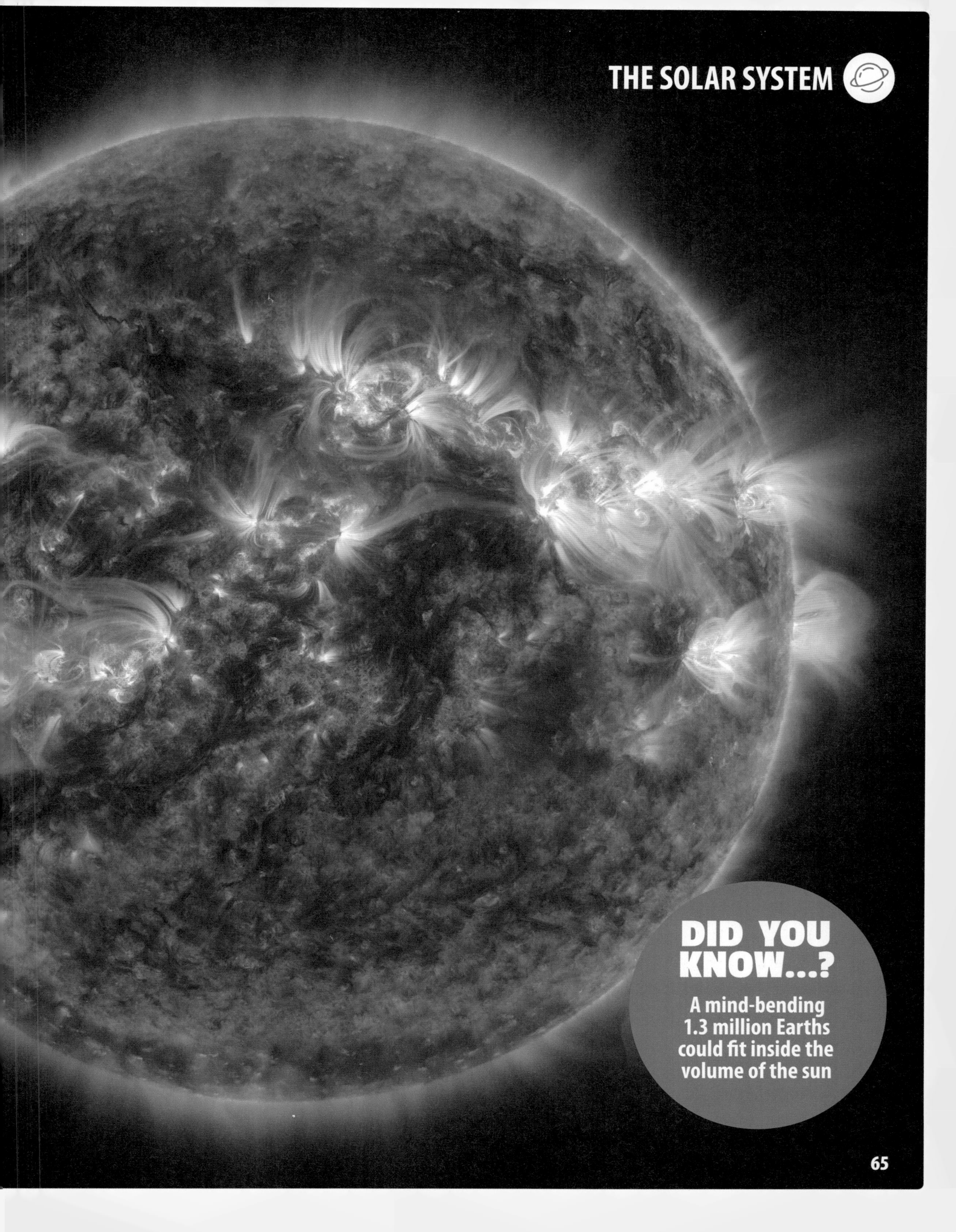

DID YOU KNOW...?

A mind-bending 1.3 million Earths could fit inside the volume of the sun

Coronal mass ejections blast plasma out into space at speeds of hundreds or thousands of miles per hour

from 900,000 F (500,000 C) to 10.8 million F (6 million C) and can even reach tens of millions of degrees when a solar flare occurs. Matter from the corona is blown off as the solar wind.

MAGNETIC FIELD

The strength of the sun's magnetic field is typically only about twice as strong as Earth's field. However, it becomes highly concentrated in small areas, reaching up to 3,000 times stronger than usual.

These kinks and twists in the magnetic field develop because the sun spins more rapidly at the equator than at the higher latitudes and because the inner parts of the sun rotate more quickly than the surface. These distortions create features ranging from sunspots to spectacular eruptions known as flares and coronal mass ejections. Flares are the most violent eruptions in the solar system, while coronal mass ejections are less violent but involve extraordinary amounts of matter – a single ejection can spout roughly 20 billion tons (18 billion metric tons) of matter into space.

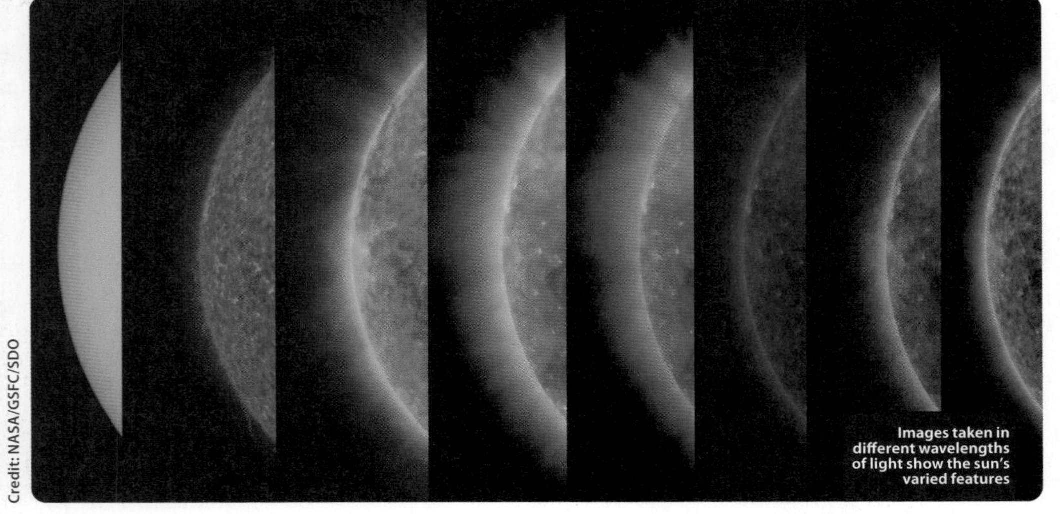

Images taken in different wavelengths of light show the sun's varied features

CHEMICAL COMPOSITION

Just like most other stars, the sun is made up mostly of hydrogen, followed by helium. Nearly all the remaining matter consists of seven other elements – oxygen, carbon, neon, nitrogen, magnesium, iron and silicon. For every 1 million atoms of hydrogen in the sun, there are 98,000 of helium, 850 of oxygen, 360 of carbon, 120 of neon, 110 of nitrogen, 40 of magnesium, 35 of iron and 35 of silicon. Still, hydrogen is the lightest of all elements, so it only accounts for roughly 72 percent of the sun's mass, while helium makes up about 26 percent.

SUNSPOTS AND SOLAR CYCLES

Sunspots are relatively cool, dark features on the sun's surface that are often roughly circular. These features emerge where dense bundles of magnetic field lines from the sun's interior break through the surface.

The number of sunspots varies as solar magnetic activity does – the change in this number, from a minimum of none to a maximum of roughly 250 sunspots or clusters of sunspots and then back to a minimum, is known as the solar cycle, and averages about 11 years long. At the end of a cycle, the magnetic field rapidly reverses its polarity.

"KINKS AND TWISTS IN THE MAGNETIC FIELD DEVELOP BECAUSE THE SUN SPINS MORE RAPIDLY AT THE EQUATOR, AND THE INNER PARTS ROTATE MORE QUICKLY THAN THE SURFACE"

Coronal loops over active regions rain plasma back down on the solar surface

Credit: NASA Goddard

Specialised solar telescopes monitor the sun's activity

Credit: NASA/JPL-Caltech/GSFC/JAXA

An illustration depicting the sun's complex magnetic field

Credit: NASA Goddard

OBSERVATION & HISTORY

Ancient cultures often modified natural rock formations or built stone monuments to mark the motions of the sun and moon, charting the seasons, creating calendars and monitoring eclipses. Many believed the sun revolved around the Earth, with ancient Greek scholar Ptolemy formalizing this "geocentric" model in 150 B.C.

Then, in 1543, Nicolaus Copernicus described a heliocentric, sun-centered model of the solar system, and in 1610, Galileo Galilei's discovery of Jupiter's moons revealed that not all heavenly bodies circled the Earth.

To learn more about how the sun and other stars work, after early observations using rockets, scientists began studying the sun from Earth orbit. NASA launched a series of eight orbiting observatories known as the Orbiting Solar Observatory between 1962 and 1971. Seven of them were successful, and analyzed the sun at ultraviolet and X-ray wavelengths and photographed the super-hot corona, among other achievements.

In 1990, NASA and the European Space Agency launched the Ulysses probe to make the first observations of its polar regions. In 2004, NASA's Genesis spacecraft returned samples of the solar wind to Earth for study. In 2007, NASA's double-spacecraft Solar Terrestrial Relations Observatory (STEREO) mission returned the first three-dimensional images of the sun. NASA lost contact with STEREO-B in 2014, which is remained out of contact except for a brief period in 2016. STEREO-A remains fully functional.

One of the most important solar missions to date has been the Solar and Heliospheric Observatory (SOHO), which was designed to study the solar wind, as well as the sun's outer layers and interior structure. It has imaged the structure of sunspots below the surface, measured the acceleration of the solar wind, discovered coronal waves and solar tornadoes, found more than 1,000 comets, and revolutionized our ability to forecast space weather.

In recent years, NASA's Solar Dynamics Observatory (SDO) – the most advanced spacecraft yet designed to study the sun – has returned never-before-seen details of material streaming outward and away from sunspots, as well as extreme close-ups of activity on the sun's surface and the first high-resolution measurements of solar flares in a broad range of extreme ultraviolet wavelengths.

There are other missions planned to observe the sun in the next few years. The European Space Agency's Solar Orbiter launched in 2018, and by 2021 will be in operational orbit around the sun. Its closest approach to the sun will be 26 million miles (43 million kilometres) – about 25 percent closer than Mercury. Solar Orbiter will look at particles, plasma and other items in an environment relatively close to the sun, before these things are modified by being transported across the solar system. The goal is to better understand the solar surface and the solar wind.

The Parker Solar Probe launched in 2018 to make an extremely close approach to the sun, getting as near as 4 million miles (6.5 million km) in 2025. It will look at the corona to learn more about how energy flows through the sun, the structure of the solar wind, and how energetic particles are transported.

25 WEIRDEST SOLAR SYSTEM FACTS

READ ON TO FIND OUT SOME OF THE STRANGEST FACTS ABOUT PLANETS, DWARF PLANETS, COMETS AND OTHER INCREDIBLE OBJECTS AROUND THE SOLAR SYSTEM

WORDS: ELIZABETH HOWELL

URANUS IS TILTED ON ITS SIDE

Uranus appears to be a featureless blue ball upon first glance, but this gas giant of the outer solar system is pretty weird upon closer inspection. First, the planet rotates on its side for reasons scientists haven't quite figured out. The most likely explanation is that it underwent some sort of one or more titanic collisions in the ancient past. In any case, the tilt makes Uranus unique among the solar system planets.

Uranus also has tenuous rings, which were confirmed when the planet passed in front of a star (from Earth's perspective) in 1977; as the star's light winked on and off repeatedly, astronomers realized there was more than just a planet blocking its starlight. More recently, astronomers spotted storms in Uranus' atmosphere several years after its closest approach to the sun, when the atmosphere would have been heated the most.

Credit: Getty Images

JUPITER'S MOON IO HAS TOWERING VOLCANIC ERUPTIONS

For those of us used to Earth's relatively inactive moon, Io's chaotic landscape may come as a huge surprise. The Jovian moon has hundreds of volcanoes and is considered the most active moon in the solar system, sending plumes up to 250 miles into its atmosphere. Some spacecraft have caught the moon erupting; the Pluto-bound New Horizons craft caught a glimpse of Io bursting when it passed by in 2007.

Io's eruptions come from the immense gravity the moon is exposed to, being nestled in Jupiter's gravitational well. The moon's insides tense up and relax as it orbits closer to, and farther from, the planet, generating enough energy for volcanic activity. Scientists are still trying to figure out how heat spreads through Io's interior, though, making it difficult to predict where the volcanoes exist using scientific models alone.

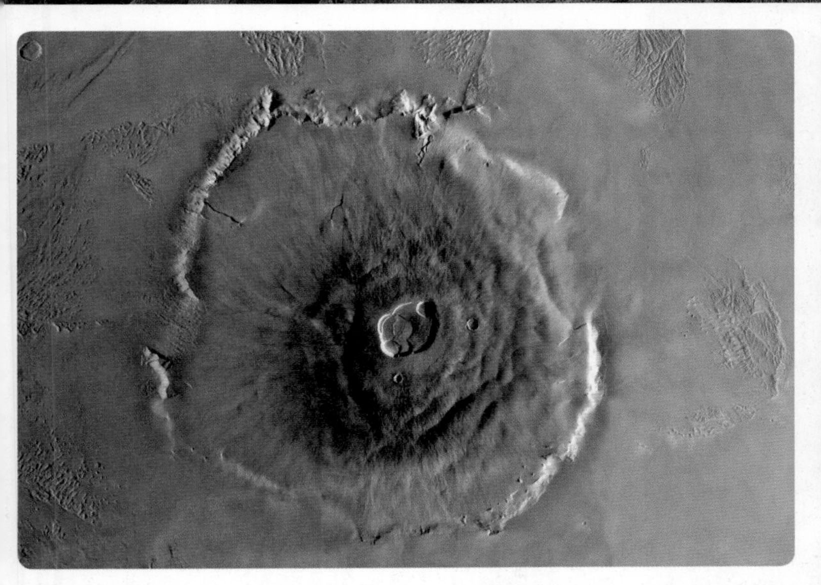

MARS HAS THE BIGGEST VOLCANO (THAT WE KNOW OF)

While Mars seems quiet now, we know that in the past something caused gigantic volcanoes to form and erupt. This includes Olympus Mons, the biggest volcano ever discovered in the solar system. At 374 miles (602 km) across, the volcano is comparable to the size of Arizona. It's 16 miles (25 km) high – triple the height of Mount Everest.

Volcanoes on Mars can grow to such immense size because gravity is much weaker on the Red Planet than it is here on Earth. But how those volcanoes came to be in the first place is not well known. There is a debate among scientists as to whether or not Mars has a global plate tectonic system and whether it is active.

MARS ALSO HAS THE LONGEST VALLEY

If you thought the Grand Canyon was big, that's nothing compared to Valles Marineris. At 2,500 miles (4,000 km) long, this immense system of Martian canyons is more than 10 times as long as the Grand Canyon on Earth. Valles Marineris escaped the notice of early Mars spacecraft (which flew over other parts of the planet) and was finally spotted by the global mapping mission Mariner 9 in 1971. And what a sight to miss – Valles Marineris is about as long as the US!

The lack of active plate tectonics on Mars makes it tough to figure out how the canyon formed. Some scientists even think that a chain of volcanoes on the other side of the planet, known as the Tharsis Ridge, somehow bent the crust from the opposite side of Mars.

VENUS HAS SUPER-POWERFUL WINDS

Venus is a hellish planet with a high-temperature, high-pressure environment on its surface. Ten of the Soviet Union's heavily shielded Venera spacecraft lasted only a few minutes on its surface when they landed there in the 1970s. But even above its surface, the planet has a bizarre environment. Scientists have found that its upper winds flow 50 times faster than the planet's rotation. The European Venus Express spacecraft (which orbited the planet between 2006 and 2014) tracked the winds over long periods and detected periodic variations. It also found that the hurricane-force winds appeared to be getting stronger over time.

THERE IS WATER ICE EVERYWHERE

Credit: NASA/JPL-Caltech/UCLA/MPS/DLR/IDA

Water ice was once considered a rare substance in space, but now we know we just weren't looking for it in the right places. In fact, water ice exists all over the solar system.

Ice is a common component of comets and asteroids, for example. But we know that not all ice is the same. Close-up examination of Comet 67P/Churyumov–Gerasimenko by the ESA's Rosetta spacecraft revealed a different kind of water ice than what is found on Earth.

That said, we've spotted water ice all over the solar system. It's in permanently shadowed craters on Mercury and the moon, although we don't know if there would be enough to support colonies in those places.

Mars also has ice at its poles, in frost and likely below the surface dust. Even smaller bodies in the solar system have ice – Jupiter's moon Europa, Saturn's moon Enceladus, and the dwarf planet Ceres, among others.

Credit: Getty Images

THERE COULD BE LIFE IN THE SOLAR SYSTEM, SOMEWHERE

So far, scientists have found no evidence that life exists elsewhere in the solar system. But as we learn more about how "extreme" microbes live in underwater volcanic vents or in frozen environments, more possibilities open up for where they could live on other planets. These aren't the aliens people once feared lived on Mars, but microbial life in the solar system is a possibility.

Microbial life is now considered so likely on Mars that scientists take special precautions to sterilize spacecraft before sending them over there. That's not the only place, though. With several icy moons scattered around the solar system, it's possible there are microbes somewhere in the oceans of Jupiter's Europa, or perhaps underneath the ice at Saturn's Enceladus, among other locations across the solar system.

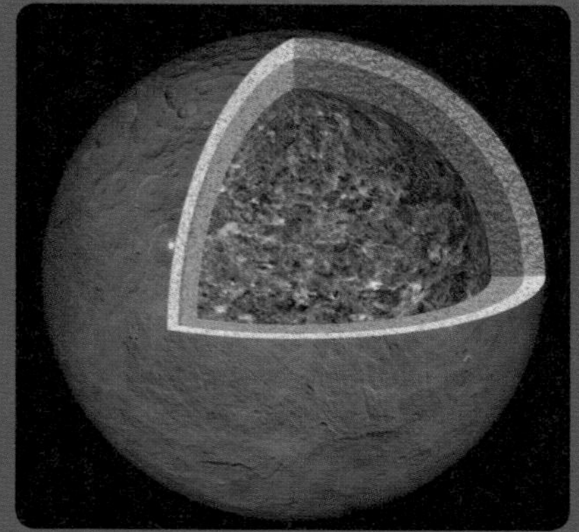

Credit: NASA/JPL-Caltech/SETI Institute

SPACECRAFT HAVE VISITED EVERY PLANET

We've been exploring space for more than 60 years, and have been lucky enough to get close-up pictures of dozens of celestial objects. Most notably, we've sent spacecraft to all of the planets in our solar system – Mercury, Venus, Earth, Mars, Jupiter, Saturn, Uranus and Neptune – as well as two dwarf planets, Pluto and Ceres.

The bulk of the flybys came from NASA's twin Voyager spacecraft, which left Earth in 1977 and are still transmitting data from beyond the solar system in interstellar space. Between them, the Voyagers clocked visits to Jupiter, Saturn, Uranus and Neptune, thanks to an opportune alignment of the outer planets.

RINGS ARE EVERYWHERE IN THE SOLAR SYSTEM

While we've known about Saturn's rings since telescopes were invented in the 1600s, it took spacecraft and more powerful telescopes built in the last 50 years to reveal more. We now know that every planet in the outer solar system – Jupiter, Saturn, Uranus and Neptune – each have ring systems. That said, rings are very different from planet to planet. Saturn's spectacular rings, which may have come from a broken-up moon, are not repeated anywhere else.

Rings aren't limited to planets, either. In 2014, astronomers discovered rings were discovered around the asteroid Chariklo. Why such a small body would have rings is a mystery, but one hypothesis is perhaps a broken-up moonlet created the fragments.

Credit: NASA/JPL/USGS

THERE ARE MOUNTAINS ON PLUTO

Pluto is a tiny world at the edge of the solar system, so at first it was thought that the dwarf planet would have a fairly uniform environment. That changed when NASA's New Horizons spacecraft flew by there in 2015, sending back pictures that altered our view of Pluto forever.

Among the astounding discoveries were icy mountains that are 11,000ft (3,300m) high, indicating that Pluto must have been geologically active as little as 100 million years ago. But geological activity requires energy, and the source of Pluto's energy is a mystery. The sun is too far away to generate enough heat for geological activity, and there are no large planets nearby that could have caused such disruption by gravity.

MERCURY IS STILL SHRINKING

For many years, scientists believed that Earth was the only tectonically active planet in the solar system. That changed after the Mercury Surface, Space Environment, Geochemistry and Ranging (MESSENGER) spacecraft did the first orbital mission at Mercury, mapping the entire planet in high definition and getting a look at the features on its surface.

In 2016, data from MESSENGER (which had crashed into Mercury as planned in April 2015) revealed cliff-like landforms known as fault scarps. Because the fault scarps are relatively small, scientists are sure that they weren't created that long ago and that the planet is still contracting 4.5 billion years after the solar system was formed.

Credit: NASA Goddard

PLUTO HAS A BIZARRE ATMOSPHERE

The observed atmosphere on Pluto broke all the predictions. Scientists saw the haze extending as high as 1,000 miles (1,600 km), rising higher above the surface than the atmosphere on Earth. As data from New Horizons flowed in, scientists analyzed the haze and discovered some surprises there, too.

Scientists found about 20 layers in Pluto's atmosphere that are both cooler and more compact than expected. This affects calculations for how quickly Pluto loses its nitrogen-rich atmosphere to space. NASA's New Horizons team found that tons of nitrogen gas escape the dwarf planet by the hour, but somehow Pluto is able to constantly resupply that lost nitrogen. The dwarf planet is likely creating more of it through geological activity.

Credit: NASA/Johns Hopkins University Applied Physics Laboratory/Southwest Research Institute

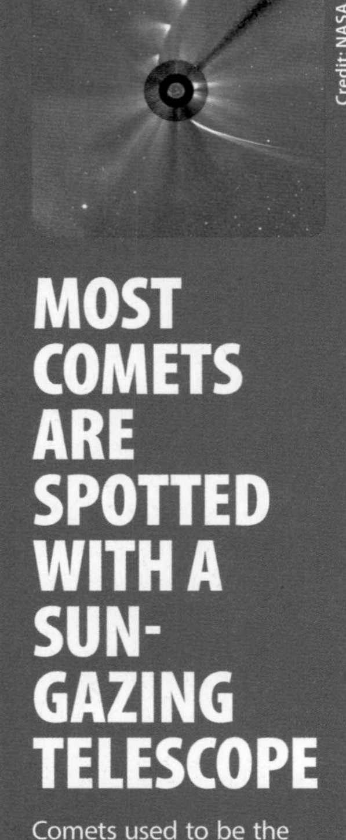
Credit: NASA Goddard

MOST COMETS ARE SPOTTED WITH A SUN-GAZING TELESCOPE

Comets used to be the province of amateur astronomers, who spent night after night scouring the skies with telescopes. While some professional observatories also made discoveries while viewing comets, that really began to change with the launch of the Solar and Heliospheric Observatory (SOHO) in 1995.

Since then, the spacecraft has found more than 2,400 comets, which is incredible considering its primary mission is to observe the sun. These comets are nicknamed "sungrazers" because they come so close to the sun. Many amateurs still participate in the search for comets by picking them out from raw SOHO images. One of SOHO's most famous observations came when it watched the breakup of the bright Comet ISON in 2013.

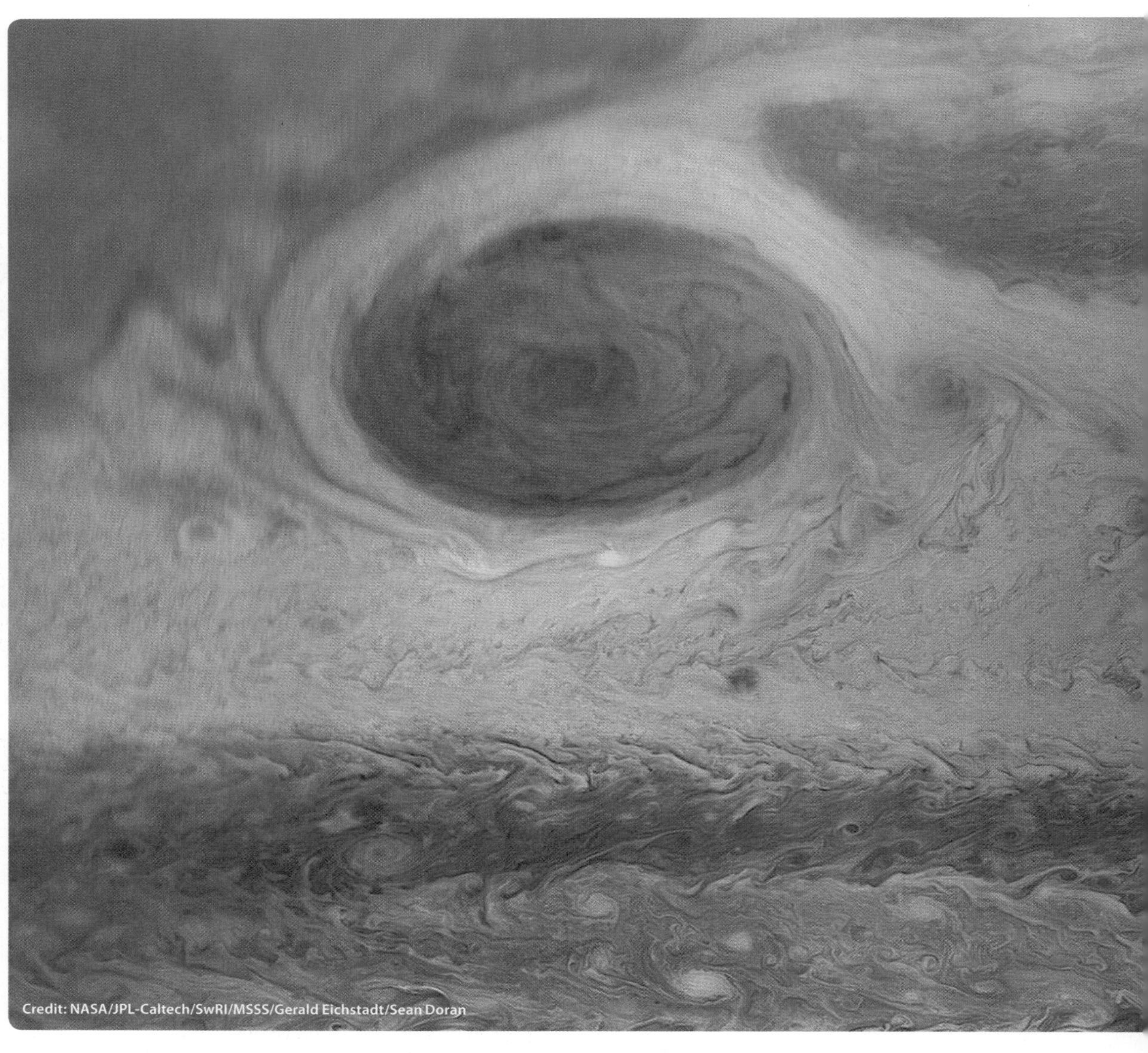
Credit: NASA/JPL-Caltech/SwRI/MSSS/Gerald Eichstadt/Sean Doran

JUPITER HAS MORE HEAVY ELEMENTS (PROPORTIONALLY) THAN THE SUN

The sun and the planets likely formed from the same cloud of hydrogen and helium gas. This would especially be true of Jupiter, a planet 317 times the size of Earth that pulled in a lot more gas than our own planet. So if that's the case, why does Jupiter have more heavy, rocky elements than the sun?

One of the leading theories is that Jupiter's atmosphere is "enriched" by the comets, asteroids and other small rocky bodies that it pulls in with its strong gravitational field. Since amateur technology has improved, several small bodies have been seen falling into Jupiter in the past decade.

Credit: NASA GSFC

JUPITER'S GREAT RED SPOT IS SHRINKING

Along with being the solar system's largest planet, Jupiter also hosts the solar system's largest storm. Known as the Great Red Spot (since it's big and ruddy-colored), it's been observed in telescopes since the 1600s. Nobody knows exactly why the storm has been raging for centuries, but in recent decades another mystery emerged: the spot is getting smaller.

In 2014, the storm was only 10,250 miles (16,500 km) across, about half of what was measured historically. The shrinkage is being monitored in professional telescopes and also by amateurs, as telescope and computer technology allow high-powered photographs at an affordable cost. Amateurs are often able to make more consistent measurements of Jupiter, because viewing time on larger, professional telescopes is limited and often split between different objects.

THERE MAY BE A HUGE PLANET AT THE EDGE OF THE SOLAR SYSTEM

In January 2015, California Institute of Technology astronomers Konstantin Batygin and Mike Brown announced – based on mathematical calculations and on simulations – that there could be a giant planet lurking far beyond Neptune. Several teams are now on the search for this theoretical "Planet Nine," which could take decades to find (if it's even out there.)

This large object, if it exists, could help explain the movements of some objects in the Kuiper Belt, an icy collection of objects beyond Neptune's orbit. Brown has already discovered several large objects in that area that in some cases rivaled or exceeded the size of Pluto. (His discoveries were one of the catalysts for changing Pluto's status from planet to dwarf planet in 2006.)

NEPTUNE RADIATES MORE HEAT THAN IT GETS FROM THE SUN

Neptune is far away from Earth, and you can bet that scientists would love to get another spacecraft out there sometime soon. Perhaps today's technology could better answer some Neptunian mysteries, such as why the blue planet is giving off more heat than it receives. It's bizarre, considering that Neptune is so far away from the faint sun.

Scientists would love to know what's going on, because it's believed that the vast temperature differential could affect weather processes on the planet. NASA estimates the temperature difference between the heat source and the cloud tops is minus 260 degrees Fahrenheit (minus 160 degrees Celsius).

Credit: NASA Goddard

EARTH'S VAN ALLEN BELTS ARE MORE BIZARRE THAN EXPECTED

Earth has bands of radiation belts surrounding our planet, known as the Van Allen belts (named after the discoverer of this phenomenon.) While we've known about the belts since the dawn of the space age, the Van Allen Probes (launched in 2012) have provided our best-ever view of them. And they've uncovered quite a few surprises along the way.

We now know that the belts expand and contract according to changes in solar activity. Sometimes the belts are very distinct, and sometimes they swell into one massive belt. An extra radiation belt (beyond the known two) was spotted in 2013. Understanding these belts helps scientists make better predictions about space weather and solar storms.

TITAN HAS A LIQUID CYCLE, BUT IT'S DEFINITELY NOT WATER

Another weird moon in Saturn's system is Titan, which hosts a liquid "cycle" that moves between the atmosphere and the surface. That sounds a lot like Earth, until you begin looking at its environment. It has lakes filled with methane and ethane, which could be reminiscent of the chemistry that occurred on Earth before life arose.

Titan also nitrogen-rich compounds known as tholins. This gives the Saturnian moon its distinctive orange color. Titan's atmosphere is so thick that radar is needed to penetrate a spacecraft's view down to the surface.

URANUS HAS A VERY BATTERED MOON

One of the most bizarre moons in the outer solar system is Miranda, which, unfortunately, we saw only once when Voyager 2 passed by it in 1986. This moon of Uranus has bizarre features on its surface, with sharp boundaries separating ridges, craters and other things. It is possible that the moon could have had tectonic activity, but how that happened on a body just 310 mi (500 km) across is a mystery.

Scientists aren't sure how the patchwork surface came about, and we likely won't be able to tell for sure until another mission gets out there. Perhaps the moon was smashed into bits and coalesced again, or maybe meteorites struck the surface and caused temporary melts in small areas.

SATURN HAS A TWO-TONE MOON

Saturn's moon Iapetus is a study in contrasts, with a very dark hemisphere and a very light hemisphere. It's unlike anything else in the solar system and has sparked speculation as to what is really going on.

Some scientists believe that particles from Phoebe (another, darker moon) may be falling on its surface. Others speculate that it's due to volcanic eruptions of hydrocarbons, creating dark patches. Cassini's flyby of Iapetus in 2007 also postulated a third theory, which is thermal segregation. Iapetus only rotates once every 79 days or so, stretching out the daily temperature cycle. This could force icy material to move into colder regions as the dark material heats up.

Credit: NASA/JPL/Space Science Institute

SATURN HAS A HEXAGONAL-SHAPED STORM

Saturn's northern hemisphere has a raging six-sided storm nicknamed "the hexagon." Why exactly it's that shape is a mystery. But what is known is that this hexagon, which shares several features in common with hurricanes, has been there for at least decades – if not hundreds of years.

Lighting conditions in Saturn's northern hemisphere began to improve in 2012, when Saturn approached its northern summer solstice. Cassini will continue observing the feature until the end of its mission in 2017, at the height of the solstice.

Credit: NASA/JPL-Caltech/SSI/Hampton University

THE SOLAR ATMOSPHERE IS MUCH HOTTER THAN THE SURFACE

Saturn's northern hemisphere has a raging six-sided storm nicknamed "the hexagon." Why exactly it's that shape is a mystery. But what is known is that this hexagon, which shares several features in common with hurricanes, has been there for at least decades – if not hundreds of years.

Lighting conditions in Saturn's northern hemisphere began to improve in 2012, when Saturn approached its northern summer solstice. Cassini will continue observing the feature until the end of its mission in 2017, at the height of the solstice.

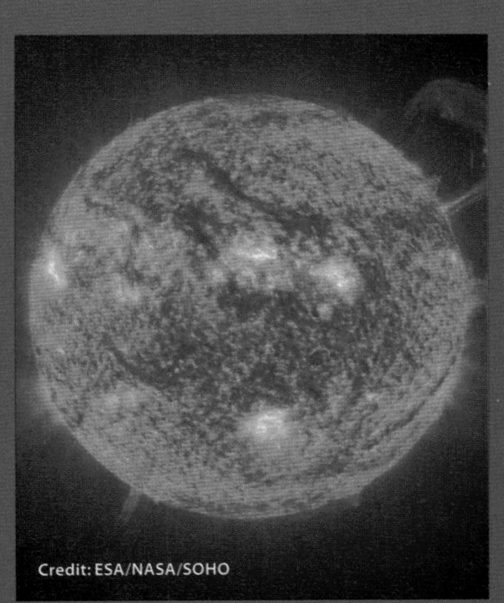
Credit: ESA/NASA/SOHO

Credit: NASA/JPL-Caltech/Univ. of Arizona

MARS HAS VARYING AMOUNTS OF METHANE IN ITS ATMOSPHERE

Methane is a substance that is produced by life (such as by microbes) or by natural processes such as volcanic activity. But why it keeps fluctuating so much on Mars is a mystery. Various telescopes and space probes have found different levels of methane on Mars over the years, making it hard to chart where this substance is coming from. It's unclear if the varying levels of methane are due to telescopic differences, or differences in the amount of methane coming from the surface.

NASA's Curiosity rover even detected a spike in methane during one Martian year that did not repeat the next, indicating whatever it saw was not seasonal. It will likely take a series of more long-term observations of Mars to fully figure out the mystery.

ORGANICS ARE COMMONPLACE IN THE SOLAR SYSTEM

Organics are molecules that are present both in life processes and in nonlife processes. While common on Earth, what's interesting is they're also in many places in the solar system. When scientists found organics on the surface of Comet 67P, for example, it bolstered the case that perhaps organics were brought to the surface of Earth by small bodies.

Organics have also been found on the surface of Mercury, on Saturn's moon Titan (which gives Titan its orange color) and on Mars. This makes them common throughout the solar system.

Credit: NASA/JPL-Caltech

COSMIC PHENOMENA

Credit: ESA/ATG medialab; ESO/S. Brunie

Credit: NASA

90

94

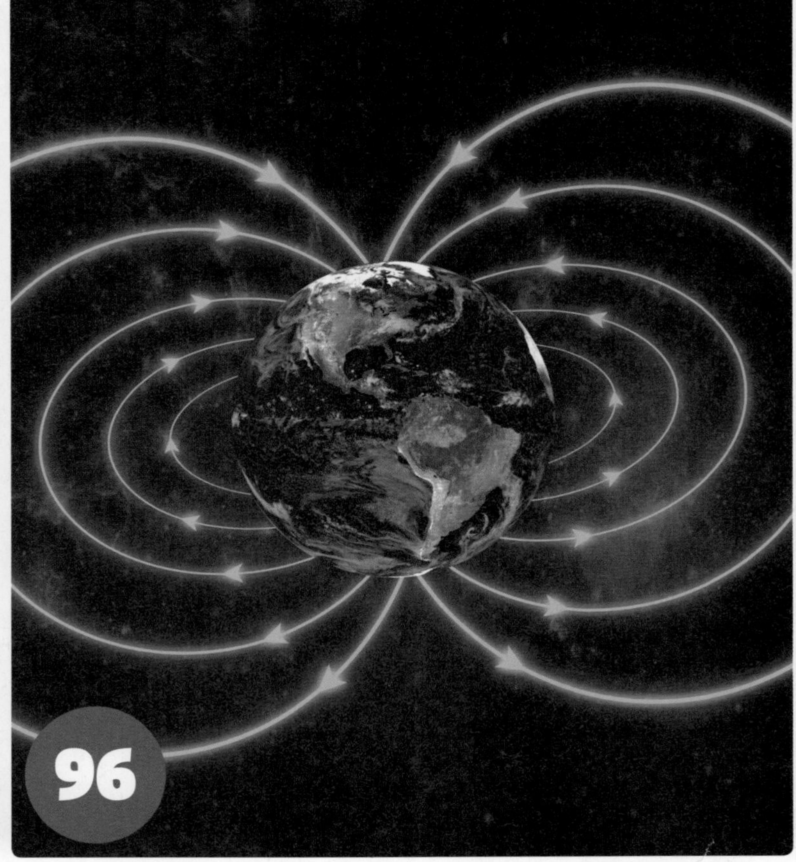
96

WEIGHING THE MILKY WAY

OUR MILKY WAY GALAXY WEIGHS AS MUCH AS 1.5 TRILLION SUNS

WORDS: MIKE WALL

We may finally know how much the Milky Way weighs. Estimates of our home galaxy's heft vary widely, from about 500 billion times the mass of our sun to 3 trillion solar masses. The number is so tough to pin down because about 85 percent of the Milky Way's mass is made up of dark matter – mysterious stuff that neither absorbs nor emits light (hence the name).

"We just can't detect dark matter directly," Laura Watkins, of the European Southern Observatory in Garching, Germany, explained in a statement. "That's what leads to the present uncertainty in the Milky Way's mass – you can't measure accurately what you can't see!"

So, Watkins and her colleagues came up with a workaround, which they reported in a new study. They measured the velocities of globular clusters, clumps of stars that orbit the Milky Way's familiar spiral disk (but are still part of our galaxy).

"The more massive a galaxy, the faster its clusters move under the pull of its gravity," study co-author N. Wyn Evans, of the University of Cambridge in England, said in the same statement.

"Most previous measurements have found the speed at which a cluster is approaching or receding from Earth – that is, the velocity along our line of sight," Evans added. "However, we were able to also measure the sideways motion of the clusters, from which the total velocity, and consequently the galactic mass, can be calculated."

That number is 1.5 trillion solar masses (within 129,000 light-years of the galactic center, to be specific). That's pretty much right in the middle of the range delineated by previous studies.

The team relied on observations by two of the most powerful astronomy tools in operation – NASA's Hubble Space Telescope and Europe's Gaia spacecraft. Gaia, which launched in December 2013, is precisely measuring the positions and motions of hundreds of millions of objects, helping researchers create the most detailed 3D map of the Milky Way ever constructed.

The team studied the motion of 46 globular clusters, 34 of which Gaia observed and 12 Hubble measured. The most distant of these stellar clumps lies about 129,000 light-years from Earth, researchers said – for perspective: The Milky Way's disk is about 100,000 light-years wide.

"85 PERCENT OF THE MILKY WAY'S MASS IS MADE UP OF MYSTERIOUS DARK MATTER"

ESA's Gaia observatory (pictured in this concept) teamed up with the Hubble Space Telescope in the study

Credit: ESA/ATG medialab; ESO/S. Brunie

A computer-generated model of the Milky Way, surrounded by the halo of globular clusters used in this study

Credit: ESA/Hubble, NASA, L. Calçada

An artist's concept of our Milky Way galaxy

THE FIRST IMAGE OF A BLACK HOLE

BLACK HOLES HAVE FINALLY BEEN DRAGGED OUT OF THE SHADOWS

WORDS: MIKE WALL

For the first time ever, humanity has photographed one of these elusive cosmic beasts, shining light on an exotic space-time realm that had long been beyond our ken.

"We have seen what we thought was unseeable," Sheperd Doeleman, of Harvard University and the Harvard-Smithsonian Center for Astrophysics, said during a press conference at the National Press Club in Washington, DC on 10 April 2019. Doeleman directs the Event Horizon Telescope (EHT) project, which captured the epic imagery. These four photos, which were unveiled at press events around the world and in a series of published papers, outline the contours of the monster black hole lurking at the heart of the elliptical galaxy M87. The imagery is mind-blowing enough in its own right. But even more significant is the trail the new results will likely blaze, researchers said.

"There's really a new field to explore," Peter Galison, a professor of physics and the history of science at Harvard, said in an EHT talk in March 2019 at the South by Southwest (SXSW) festival in Austin, Texas. "And that's ultimately what's so exciting about this." Galison, who co-founded Harvard's interdisciplinary Black Hole Initiative (BHI), compared the imagery's potential impact to that of the drawings made by English scientist Robert Hooke in the 1600s. These illustrations showed people what insects and plants look like through a microscope. "It opened a world," Galison said of Hooke's work.

Black hole mergers produce gravitational waves, which were first detected by the LIGO experiment in 2016

Credit: Getty

A TELESCOPE THE SIZE OF EARTH

The EHT is a consortium of more than 200 scientists that has been in the works for about two decades. It's a truly international endeavor; funding over the years has come from the US National Science Foundation and many other organizations in countries around the world.

The ambitious project, like the 1997 horror sci-fi film, takes its name from a black hole's famed point of no return – the boundary beyond which nothing, not even light, can escape the massive object's gravitational clutches.

"The event horizon is the ultimate prison wall," BHI founding director Avi Loeb, the chair of Harvard's astronomy department, told Space.com (Loeb is not part of the EHT team). "Once you're in, you can never get out."

It's therefore impossible to photograph the interior of a black hole, unless you somehow manage to get in there yourself (you and your pictures couldn't make it back to the outside world again, of course). So, the EHT images the event horizon, mapping out the black hole's dark silhouette. The disk of fast-moving gas swirling around and into black holes emits lots of radiation, so such silhouettes stand out.

"We're looking for the loss of photons," EHT science council member Dan Marrone, an associate professor of astronomy at the University of Arizona, told Space.com.

The project has been scrutinizing two black holes – the M87 behemoth, which harbors about 6.5 billion times the mass of Earth's Sun, and our own Milky Way galaxy's central black hole, known as Sagittarius A*. This latter object, while still a supermassive black hole, is a runt compared to M87's beast, containing a mere 4.3 million solar masses.

The Event Horizon Telescope captured our first direct visual evidence of a supermassive black hole

Credit: EHT Collaboration

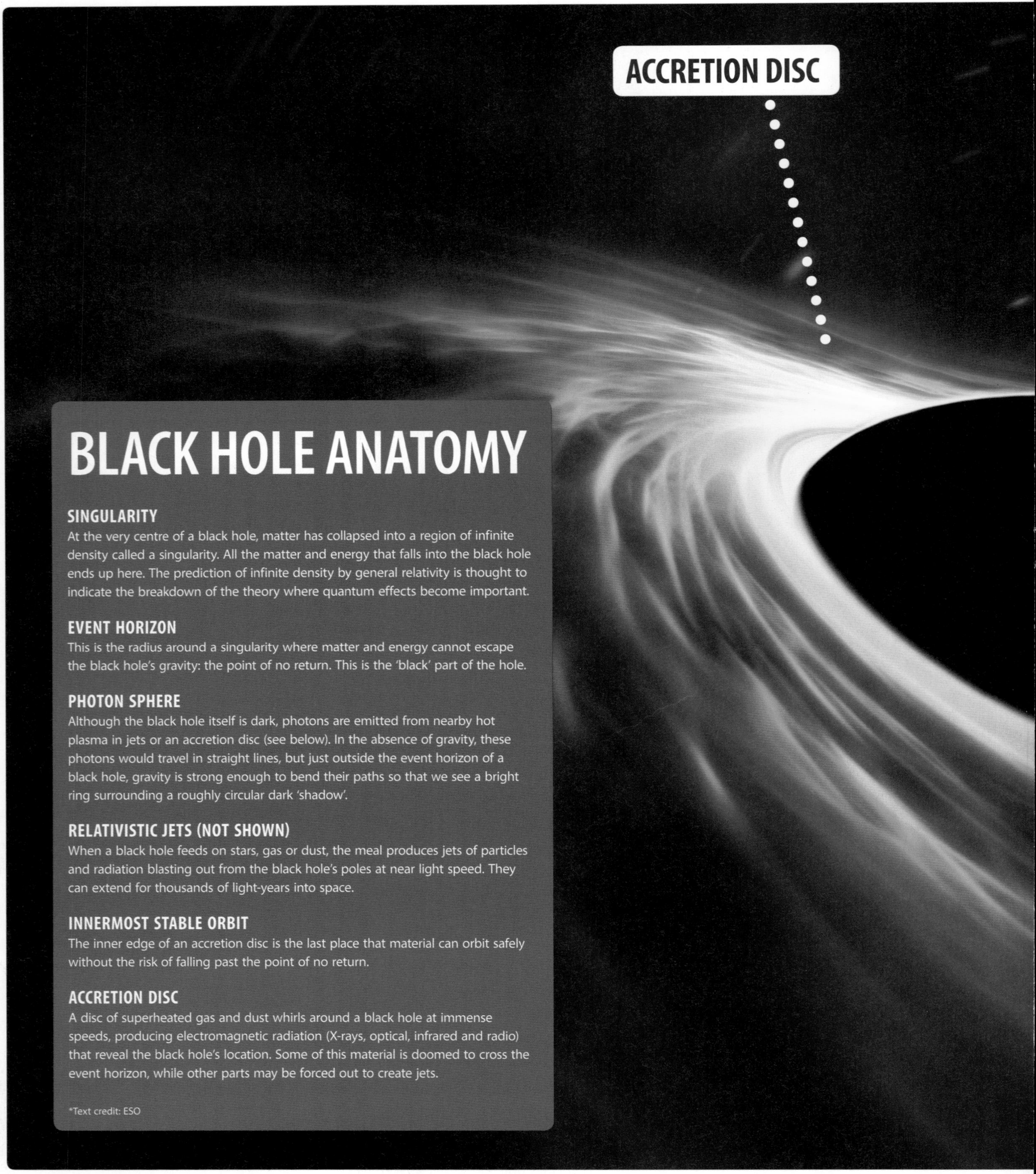

ACCRETION DISC

BLACK HOLE ANATOMY

SINGULARITY
At the very centre of a black hole, matter has collapsed into a region of infinite density called a singularity. All the matter and energy that falls into the black hole ends up here. The prediction of infinite density by general relativity is thought to indicate the breakdown of the theory where quantum effects become important.

EVENT HORIZON
This is the radius around a singularity where matter and energy cannot escape the black hole's gravity: the point of no return. This is the 'black' part of the hole.

PHOTON SPHERE
Although the black hole itself is dark, photons are emitted from nearby hot plasma in jets or an accretion disc (see below). In the absence of gravity, these photons would travel in straight lines, but just outside the event horizon of a black hole, gravity is strong enough to bend their paths so that we see a bright ring surrounding a roughly circular dark 'shadow'.

RELATIVISTIC JETS (NOT SHOWN)
When a black hole feeds on stars, gas or dust, the meal produces jets of particles and radiation blasting out from the black hole's poles at near light speed. They can extend for thousands of light-years into space.

INNERMOST STABLE ORBIT
The inner edge of an accretion disc is the last place that material can orbit safely without the risk of falling past the point of no return.

ACCRETION DISC
A disc of superheated gas and dust whirls around a black hole at immense speeds, producing electromagnetic radiation (X-rays, optical, infrared and radio) that reveal the black hole's location. Some of this material is doomed to cross the event horizon, while other parts may be forced out to create jets.

*Text credit: ESO

Both of these objects are tough targets because of their immense distance from Earth. Sagittarius A* lies about 26,000 light-years from us, and M87's black hole is a whopping 53.5 million light-years away.

From our perspective, Sagittarius A*'s event horizon "is so small that it's the equivalent of seeing an orange on the Moon or being able to read the newspaper in Los Angeles while you're sitting in New York City," Doeleman said during the SXSW event last month.

No single telescope on Earth can make that observation, so Doeleman and the rest of the EHT team had to get creative. The researchers have linked up radio telescopes in Arizona, Spain, Mexico, Antarctica and other places around the world, forming a virtual instrument the size of Earth.

SO MUCH DATA

The EHT team has used this megascope to study the two supermassive black holes for two weeklong stretches to date – once in April 2017 and again the following year. The new imagery comes from the first observing run.

There are good reasons why it's taken two years for the project's first result to come out. For one thing, each night of observing generated about 1 petabyte of data, resulting in such a haul that the team has to move its information from place to place the old-fashioned way.

"There's no way that we can transfer this data through the internet," EHT project scientist Dimitrios Psaltis, an astronomy professor at the University of Arizona, said at the SXSW event. "So, what we actually do is, we take our hard drives and we FedEx them from place to place. This is much faster than any cable that you can ever find."

This slows and complicates analysis, of course. Data from the EHT scope near the South Pole, for example, couldn't get off Antarctica until December 2017, when it was warm enough for planes to go in and out, Marrone said.

Correlating and calibrating the data was also tricky, he added. And the team took great care with this work, given the momentous nature of the find. "If you're going to come with a big claim of imaging a black hole, you have to have big evidence, very strong evidence," Doeleman said at the SXSW explainer event.

"And on our project, we often think that people like [Albert] Einstein, [Arthur] Eddington [and Karl] Schwarzschild are kind of looking over our shoulders," he added, referring to physicists who helped pioneer our understanding of black holes. "And when you have luminaries kind of virtually checking your work, you really want to get it right."

WHAT IT ALL MEANS

The EHT project has two main goals, Psaltis said: to image an event horizon for the first time ever and to help determine if Einstein's theory of general relativity needs any revisions.

Before Einstein came along, gravity was generally regarded as a mysterious force at a distance. But general relativity describes it as the warping of space-time: Massive objects such as planets, stars and black holes create a sort of sag in space-time, much as a bowling ball would if placed on a trampoline. Nearby objects follow this curve and get funneled toward the central mass.

General relativity has held up incredibly well over the century since its introduction, passing every test that scientists have thrown at it. But the EHT's observations provide another trial, in an extreme realm where predictions may not match reality. That's because astronomers can calculate the expected size and shape of an event horizon using general relativity, Psaltis explained.

If the observed silhouette matches the theory-informed simulations, "then Einstein was 100% right," Psaltis said. "If the answer is no, then we have to tweak his theory in order to make it work with experiments. This is how science goes." And we learned at this event that no tweaks are needed, at least at the moment: EHT's M87 observations are consistent with general relativity, team members said. Namely, the event horizon is nearly circular and is the "right" size for a black hole of that immense mass.

"I have to admit, I was a little stunned that it matched so closely the predictions that we had made," EHT team member Avery Broderick, of the University of Waterloo and the Perimeter Institute for Theoretical Physics in Canada, said during the EHT news conference.

All large galaxies are thought to have a supermassive black hole at their centre

Credit: Thinkstock

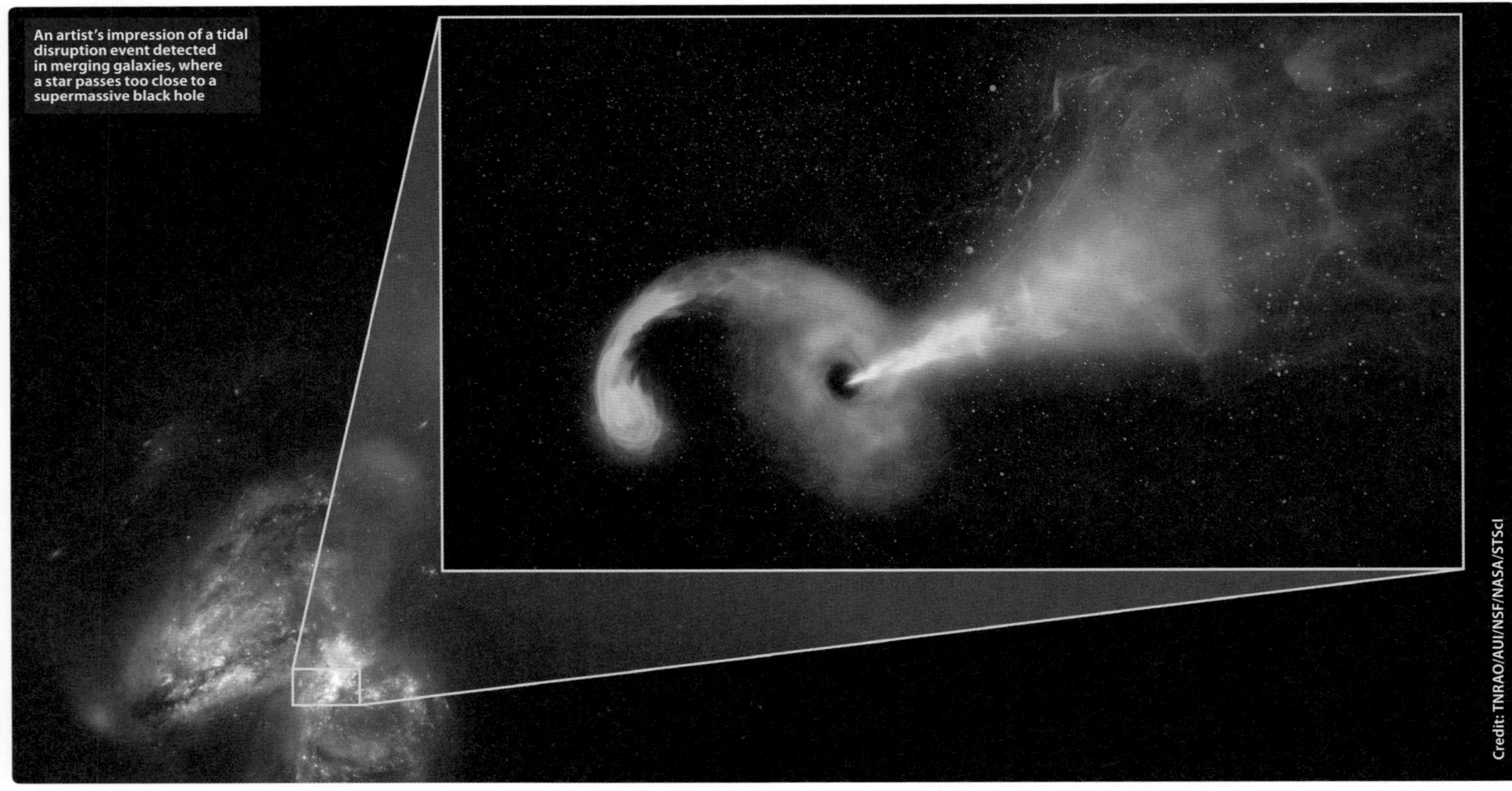

An artist's impression of a tidal disruption event detected in merging galaxies, where a star passes too close to a supermassive black hole

Credit: TNRAO/AUI/NSF/NASA/STScI

A black hole's event horizon is essentially the point of no return. The name refers to the impossibility of witnessing any event taking place inside that border, the horizon beyond which one cannot see

Such ground-truthing is vital to the scientific process, of course. Indeed, providing better information to feed into theories and simulations will likely be one of the EHT's biggest contributions, Loeb said.

"Doing physics is a dialog with nature," he said. "We test our ideas by comparing them to experiments; experimental data is crucial."

The new results should also help scientists get a better handle on black holes, he and other researchers said. For example, EHT imagery will likely shine significant light on how gas spirals down into a black hole's maw. This accretion process, which can lead to the generation of powerful jets of radiation, is poorly understood, Loeb said at the press conference.

In addition, the shape of an event horizon can reveal whether a black hole is spinning or not, said Fiona Harrison of the California Institute of Technology, the principal investigator of NASA's black-hole-studying Nuclear Spectroscopic Telescope Array (NuSTAR) mission.

"We've inferred the spin of black holes indirectly," Harrison, who's not part of the EHT team, told Space.com. EHT imagery provides "a direct test, which is very exciting," she added.

EHT's data revealed the M87 black hole is spinning clockwise, team members said.

The project should also show how matter is distributed around a black hole, and EHT observations could eventually teach astronomers a great deal about how supermassive black holes shape the evolution of their host galaxies over long time scales, Harrison told us.

EHT's results also mesh well with those of the Laser Interferometer Gravitational-Wave Observatory (LIGO), which has detected the space-time ripples generated by mergers involving black holes just a few dozen times more massive than the Sun.

"Despite varying across a factor of billion in mass, known black holes are all consistent with a single description," Broderick said at the press conference. "Black holes big and small are analogous in important ways. What we learn from one [type] necessarily applies to the other."

And in case you're wondering about Sagittarius A*: The EHT team hopes to get imagery of that supermassive black hole soon, Doeleman said. The researchers looked at M87 first, and it's a bit easier to resolve than Sagittarius A* because it's less variable over short timescales, he explained.

DID YOU KNOW...?

The largest supermassive black hole scientists have discovered so far has a mass of around 17 billion suns

A NEW PERSPECTIVE?

Then there's the broader appeal of the newly released imagery – how it speaks to those of us who aren't astrophysicists.

The contributions in this arena could be significant, EHT team members and outside scientists said. Photos can change the way we think about ourselves and our place in the universe, Marrone noted, citing the famous "Earthrise" photo taken by Apollo 8 astronaut Bill Anders in December 1968. This image, which gave the masses a glimpse of our planet as it really is – a lonely outpost of life in an infinite sea of darkness – is widely credited with helping to spur the environmental movement.

Seeing a black hole in real life – or its silhouette, anyway – "is the stuff of science fiction," Harrison said. And we've still only seen just the ambitious project's first few photos, she added: "They're only going to get better."

THE BRIGHTEST QUASARS

THESE POWERFUL DYNAMOS HAVE FASCINATED ASTRONOMERS SINCE THEIR DISCOVERY HALF A CENTURY AGO

WORDS: SARAH LEWIN

In January 2019, scientists announced the discovery of the energetic core of a distant galaxy that shatters the record for the brightest object in the early universe, blazing with the light equivalent to 600 trillion suns.

Researchers identified the object – a black-hole-powered object called a quasar, among the universe's brightest inhabitants – because of a chance alignment with a dim galaxy closer to Earth that magnified its light.

The quasar is 12.8 billion light-years away, and it shines at the heart of a forming galaxy during an early part of the universe's history called the epoch of reionization, when the first stars and galaxies began to burn away a haze of neutral hydrogen across the cosmos. Researchers announced its discovery at the January 2019 American Astronomical Society's winter meeting in Seattle.

"That's something we have been looking for for a long time," Xiaohui Fan, a researcher at the University of Arizona and lead author on the new work, said in a statement from the Hubble Space Telescope team. "We don't expect to find many quasars brighter than that in the whole observable universe!"

Several powerful ground-based telescopes and the Hubble Space Telescope pooled their observations of the object, now designated J043947.08+163415.7, to learn more about it. The quasar gets its brightness from a supermassive black hole: material from a disk of gas surrounding the black hole falls in, leading to blasts of energy at many different wavelengths, according to the statement. The quasar likely blazed when the universe was less than a billion years old, but some of its light is only now reaching Earth. According to the new observations, the black hole powering this quasar is several hundred million times the mass of the sun.

Despite its intense brightness, the distance to the quasar is so great that it wouldn't have been visible, if not for a lucky trick of positioning. Through a process called gravitational lensing, light from the quasar has bent around a galaxy in between the object and Earth, magnifying our view: the quasar appears three times as large and 50 times as bright as it would have otherwise, researchers said in the statement. And it was only observed at all because the intervening galaxy was dim enough to not drown out the light from the ultra-distant quasar.

Learning more about this quasar, which also appears to be producing 10,000 stars per year, can teach researchers more about this distant but pivotal time in history, when the first stars and galaxies were kindling and shaping the universe to what we know today. Even more telescopes are joining the search to try and discern more about the system.

"This detection is a surprising and major discovery; for decades we thought that these lensed quasars in the early universe should be very common, but this is the first of its kind that we have found," Fabio Pacucci, a researcher at Yale University, a co-author on the work and lead author on a follow-up paper about the quasar, said in a statement from the Keck Observatory. "It gives us a clue on how to search for 'phantom quasars' – sources that are out there, but cannot be really detected yet.

"Our study predicts that we might be missing a substantial fraction of these 'phantom quasars,'" Paucci added. "If they are indeed numerous, it would revolutionize our idea of what happened right after the Big Bang, and even change our view of how these cosmic monsters grew in mass."

"THE QUASAR LIKELY BLAZED WHEN THE UNIVERSE WAS LESS THAN A BILLION YEARS OLD"

An illustration of J043947.08+163415.7, the brightest quasar discovered in the early universe so far

DID YOU KNOW...?

The term "quasar" stands for "quasi-stellar radio source", and was first used in 1964

An artist's impression of a quasar

GAMMA-RAY BURSTS

THE MOST POWERFUL EXPLOSIONS IN THE UNIVERSE ARE INCREDIBLY STRANGE

WORDS: MIKE WALL

There's a weird mix of order and chaos in the light blasted out by gamma-ray bursts (GRBs), brief but intense outbursts are associated with black hole formation.

Research published in January 2019 shows that GRB photons tend to be polarized – that is, most of them oscillate in the same direction. But, surprisingly, this direction changes over time.

"The results show that, as the explosion takes place, something happens which causes the photons to be emitted with a different polarization direction," Merlin Kole, one of the new study's lead authors, said in a statement. "What this could be, we really don't know," he added.

The most powerful GRBs are unleashed when massive stars go hypernova – an especially intense type of supernova – and then collapse to form black holes. These black holes emit jets of incredibly fast-moving material along their axes of rotation. Scientists think GRB radiation is produced within these narrow relativistic jets, but exactly how that happens is unclear. More information about GRB light could help – and that's where the new study comes in.

Kole and colleagues analyzed data gathered by an instrument called POLAR, which launched to Earth orbit aboard China's Tiangong-2 space lab. As its name suggests, the instrument was designed to gauge the polarity of GRB light.

POLAR detected 55 GRBs during its operational life. For the new study, the researchers analyzed five of the most powerful bursts. They delved especially deeply into one 9-second-long GRB, partitioning it into roughly 2-second-long "slices." It was this work that revealed the surprising polarity shift.

"We now want to build POLAR-2, which is bigger and more precise," Produit said. "With that, we can dig deeper into these chaotic processes, to finally discover the source of the gamma rays and unravel the mysteries of these highly energetic physical processes."

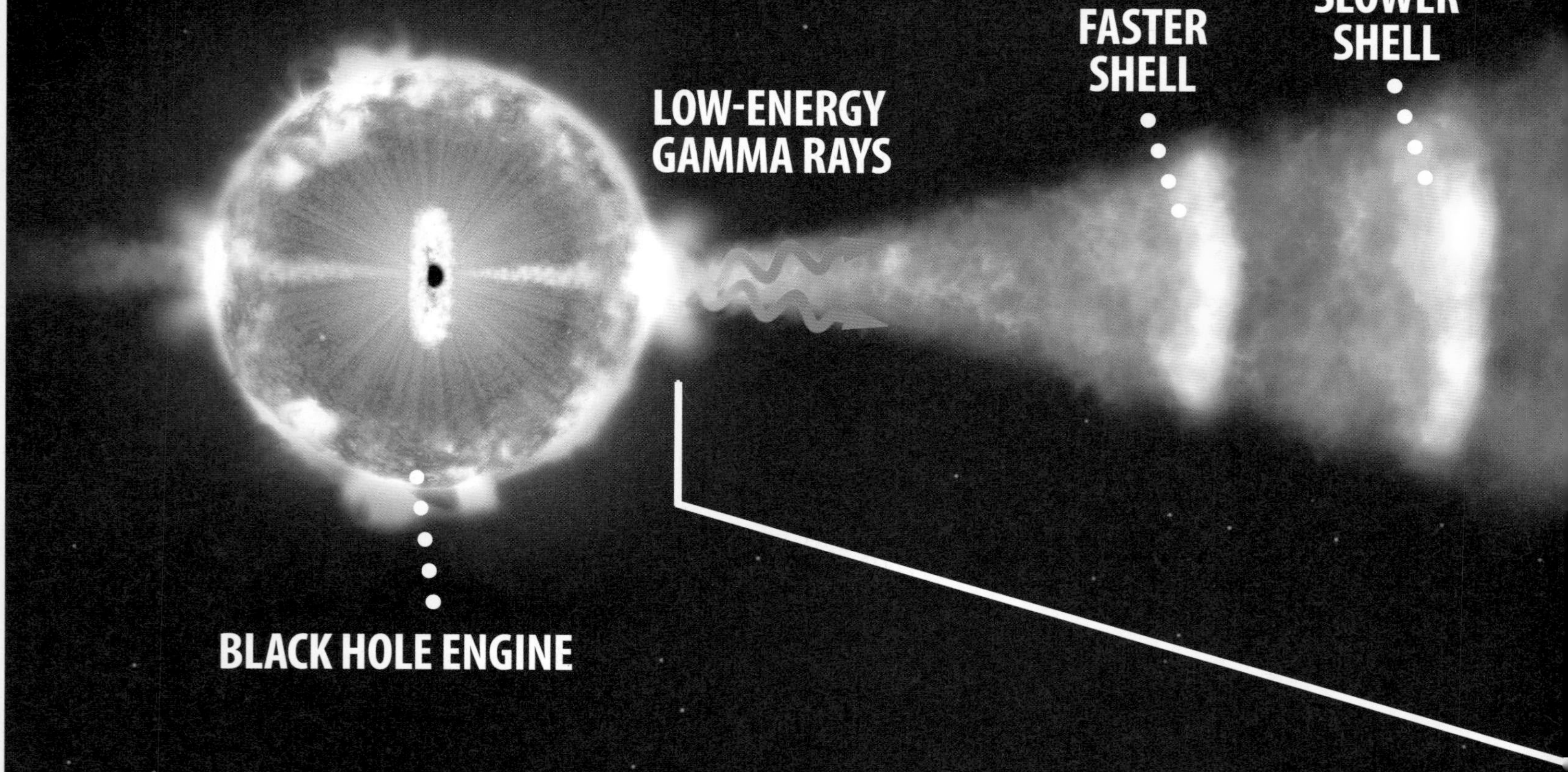

FASTER SHELL

SLOWER SHELL

LOW-ENERGY GAMMA RAYS

BLACK HOLE ENGINE

PROMPT EMISSION

"THE MOST POWERFUL GRBS ARE UNLEASHED WHEN MASSIVE STARS GO HYPERNOVA, AND THEN COLLAPSE TO FORM BLACK HOLES"

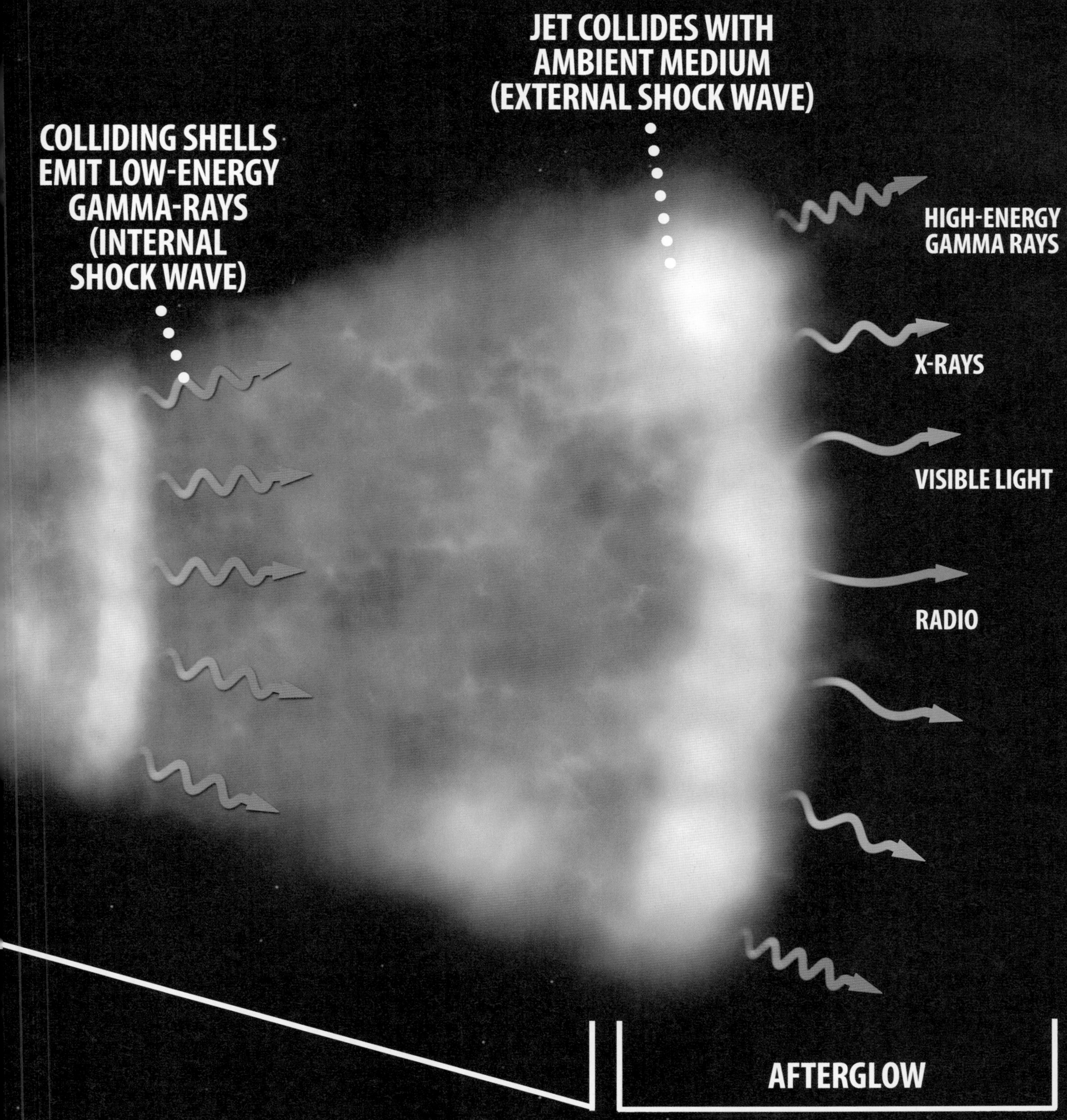

JET COLLIDES WITH AMBIENT MEDIUM (EXTERNAL SHOCK WAVE)

COLLIDING SHELLS EMIT LOW-ENERGY GAMMA-RAYS (INTERNAL SHOCK WAVE)

HIGH-ENERGY GAMMA RAYS

X-RAYS

VISIBLE LIGHT

RADIO

AFTERGLOW

8 BAFFLING ASTRONOMY MYSTERIES

THE UNIVERSE HAS BEEN AROUND FOR ROUGHLY 13.8 BILLION YEARS, BUT IT STILL HOLDS MANY MYSTERIES THAT CONTINUE TO PERPLEX ASTRONOMERS TO THIS DAY

WORDS: SPACE.COM STAFF

Ranging from dark energy to cosmic rays to the uniqueness of our own solar system, there is no shortage of cosmic oddities.

The journal Science summarized some of the most bewildering questions being asked by leading astronomers today. In no particular order, here are eight of the most enduring mysteries in astronomy

WHAT IS DARK ENERGY?

Dark energy is thought to be the enigmatic force that is pulling the cosmos apart at ever-increasing speeds, and is used by astronomers to explain the universe's accelerated expansion. This elusive force has yet to be directly detected, but dark energy is thought to make up an estimated 73 percent of the universe.

Galaxy cluster Abell 1689 with the mass distribution of the dark matter indicated in purple

Credit: NASA, ESA, E. Jullo (JPL/LAM), P. Natarajan (Yale) and J-P. Kneib (LAM)

WHY IS THE SOLAR SYSTEM SO BIZARRE?

As alien planets around other stars are discovered, astronomers have tried to tackle and understand how our own solar system came to be. The differences in the planets within our solar system have no easy explanation, and scientists are studying how planets are formed in hopes of better grasping the unique characteristics of our solar system. This research could get a boost from the hunt for alien worlds, particularly if patterns arise in their observations of extrasolar planetary systems.

It may be normal to us, but our solar system seems to be unusual compared to the exoplanet systems we've discovered so far

HOW HOT IS DARK MATTER?

Dark matter is an invisible mass that is thought to make up about 23 percent of the universe. Dark matter has mass but cannot be seen, so scientists infer its presence based on the gravitational pull it exerts on regular matter. Researchers remain curious about the properties of dark matter, such as whether it is icy cold as many theories predict, or if it is warmer.

An artist's impression of the predicted dark matter halo surrounding the Milky Way

Credit: ESO/L. Calçada

WHERE ARE THE MISSING BARYONS?

Dark energy and dark matter combine to occupy approximately 95 percent of the universe, with regular matter making up the remaining 5 percent. But, researchers have been puzzled to find that more than half of this regular matter is missing. This missing matter is called baryonic matter, and it is composed of particles such as protons and electrons that make up majority of the mass of the universe's visible matter. Some astrophysicists suspect that missing baryonic matter may be found between galaxies, in material known as warm-hot intergalactic medium, but the universe's missing baryons remain a hotly debated topic.

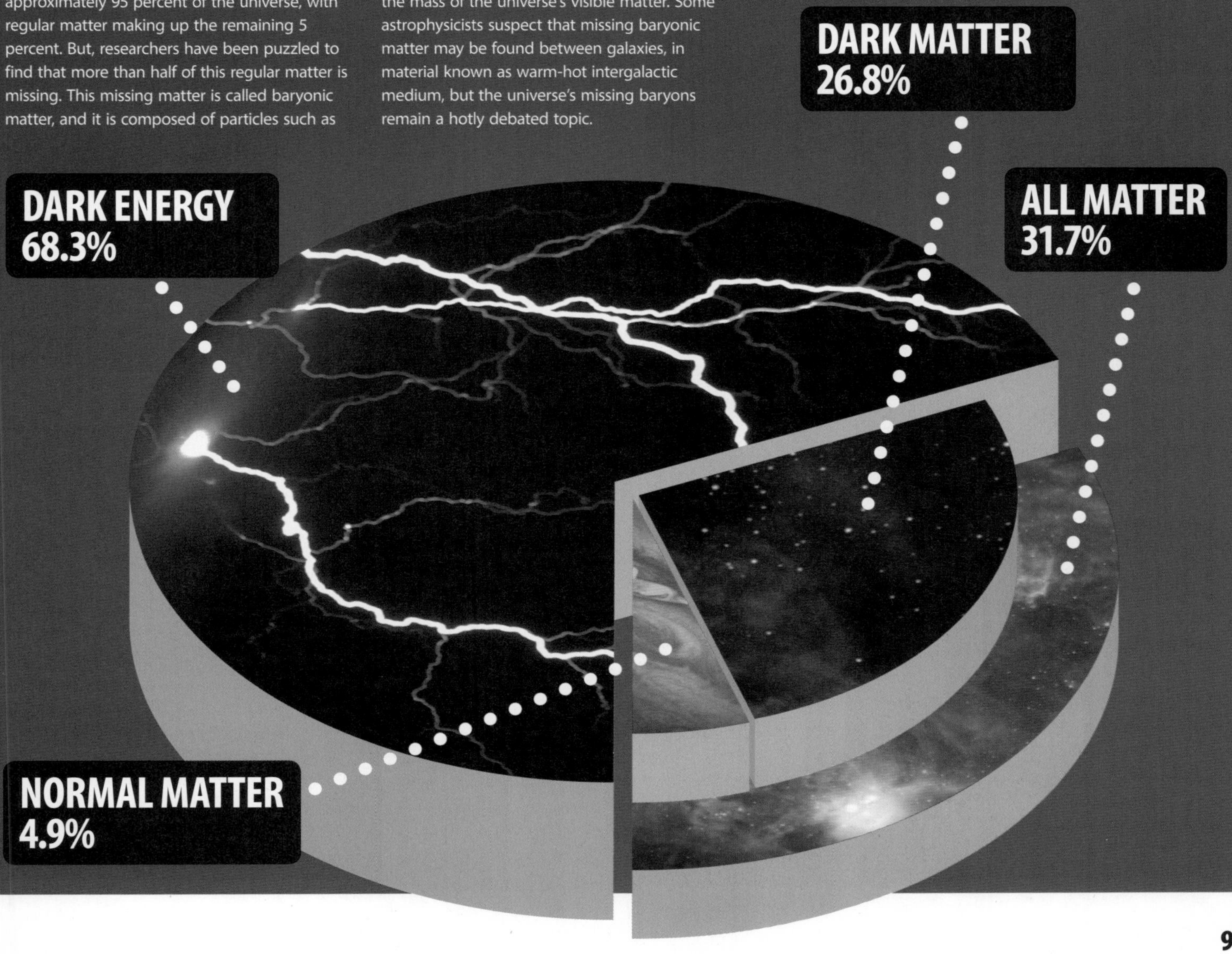

DARK MATTER
26.8%

DARK ENERGY
68.3%

ALL MATTER
31.7%

NORMAL MATTER
4.9%

HOW DO STARS EXPLODE?

When massive stars run out of fuel, they end their lives in gigantic explosions called supernovas. These spectacular blasts are so bright they can briefly outshine entire galaxies. Extensive research and modern technologies have illuminated many details about supernovas, but how these massive explosions occur is still a mystery. Scientists are keen to understand the mechanics of these stellar blasts, including what happens inside a star before it ignites as a supernova.

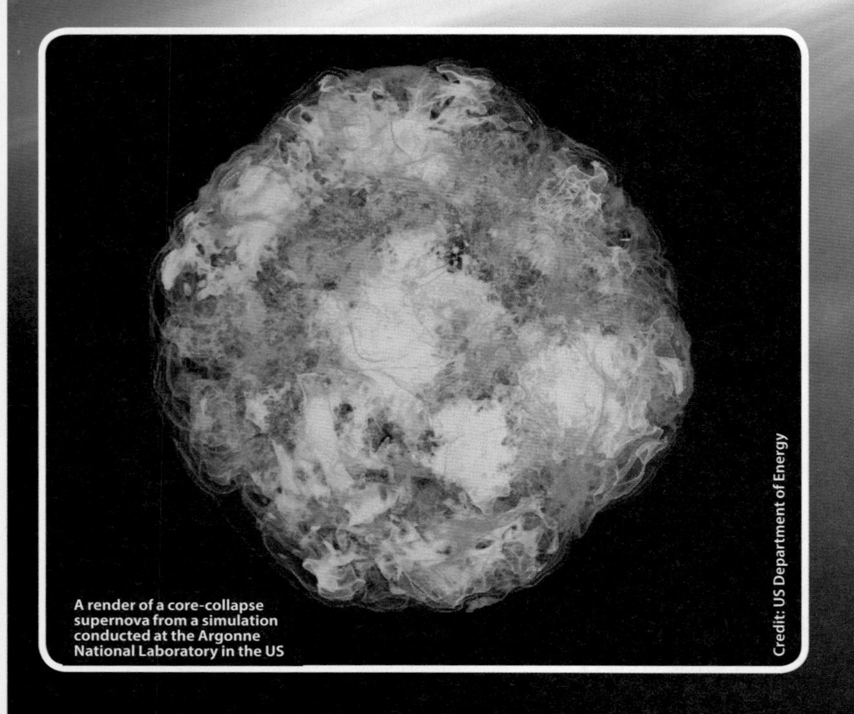

A render of a core-collapse supernova from a simulation conducted at the Argonne National Laboratory in the US

Credit: US Department of Energy

Credit: Getty

An artist's impression of a supernova event

WHAT'S THE SOURCE OF THE MOST ENERGETIC COSMIC RAYS?

Cosmic rays are highly energetic particles that flow into our solar system from deep in outer space, but the actual origin of these charged subatomic particles has perplexed astronomers for about a century. The most energetic cosmic rays are extraordinarily strong, with energies up to 100 million times greater than particles that have been produced in manmade colliders. Over the years, astronomers have attempted to explain where cosmic rays originate before flowing into the solar system, but their source has proven to be an enduring astronomical mystery.

Cosmic rays are atom fragments that rain down on the Earth from outside of the solar system

Credit: Getty

WHY IS THE SUN'S CORONA SO HOT?

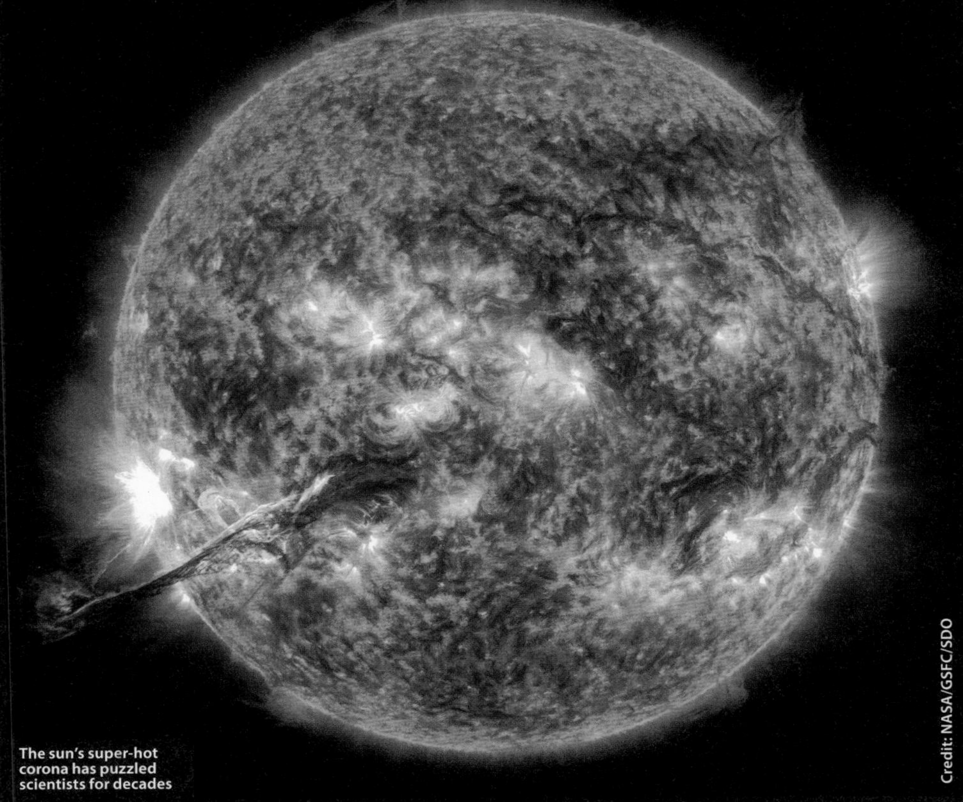

The sun's super-hot corona has puzzled scientists for decades

Credit: NASA/GSFC/SDO

The sun's corona is its ultra-hot outer atmosphere, where temperatures can reach up to a staggering 10.8 million degrees Fahrenheit (6 million degrees Celsius). Solar physicists have been puzzled by how the sun reheats its corona, but research points to a link between energy beneath the visible surface, and processes in the sun's magnetic field. But the mechanics behind coronal heating are still unknown.

WHAT RE-IONISED THE UNIVERSE?

The broadly accepted Big Bang model for the origin of the universe states that the cosmos began as a hot, dense point approximately 13.8 billion years ago. The early universe is thought to have been a dynamic place, and about 13 billion years ago, it underwent a so-called age of re-ionization. During this period, the universe's fog of hydrogen gas was clearing and becoming translucent to ultraviolet light for the first time. Scientists have long been puzzled over what caused this re-ionization to occur.

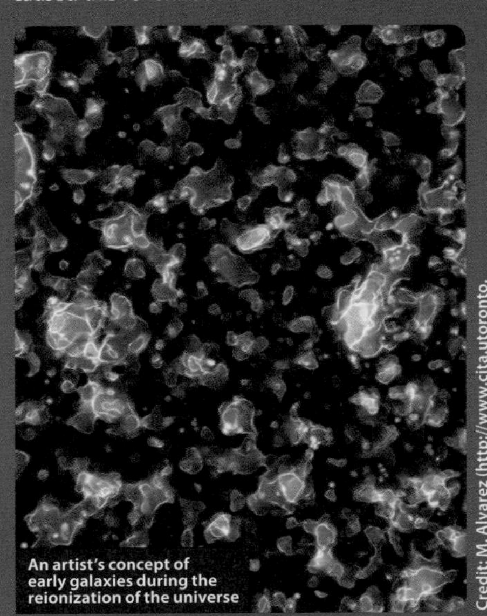

An artist's concept of early galaxies during the reionization of the universe

Credit: M. Alvarez (http://www.cita.utoronto.ca/~malvarez), R. Kaehler, and T. Abel/ESO

THE AURORA BOREALIS

WHAT CAUSES THE NORTHERN LIGHTS AND WHERE CAN THEY BE SEEN?

WORDS: SPACE.COM STAFF

At the center of our solar system lies the sun, the yellow star that sustains life on our planet. The sun's many magnetic fields distort and twist as our parent star rotates on its axis. When these fields become knotted together, they burst and create so-called sunspots. Usually, these sunspots occur in pairs; the largest can be several times the size of Earth's diameter.

As the temperature on the sun's surface rises and falls, the plasma of our star boils and bubbles. Particles escape from the star from the sunspot regions on the surface, hurtling particles of plasma, known as solar wind, out into space. It takes these winds around 40 hours to reach Earth. When they do, they can cause the dramatic displays known as the aurora borealis.

SUNSPOTS AND CYCLES

The sunspots and solar storms that cause the most magnificent displays of the northern lights occur roughly every 11 years. The solar cycle peaked in 2013, but it was the weakest solar maximum in a century. "This solar cycle continues to rank among the weakest on record," Ron Turner of Analytic Services, Inc. who serves as a Senior Science Advisor to NASA's Innovative Advanced Concepts program, said in a statement.

Since record-keeping of the ebb and flow of the sun's activity began in 1749, there have been 22 full cycles. Researchers monitor space weather events because they have the potential to affect spacecraft in orbit, knock out power grids and communications infrastructure on Earth, and amp up normal displays of the northern and southern lights. Scientists are also investigating how fluctuations in the sun's activity affect weather on our planet.

PARTICLES AND POLAR ATTRACTION

Earth is constantly bombarded with debris, radiation and other magnetic waves from space that could threaten the future of life as we know it. Most of the time, the planet's own magnetic field does an excellent job of deflecting these potentially harmful rays and particles, including those from the sun. Particles discharged from the sun travel 93 million miles (around 150 million km) toward Earth before they are drawn irresistibly toward the magnetic north and south poles. As the particles pass through the Earth's magnetic shield, they mingle with atoms and molecules of oxygen, nitrogen and other elements that result in the dazzling display of lights across the sky.

The auroras in Earth's Northern Hemisphere are called the aurora borealis. Their southern counterpart, which light up the Antarctic skies in the Southern Hemisphere, are known as the aurora australis.

WHAT CAUSES THE COLORS?

The colors most often associated with the aurora borealis are pink, green, yellow, blue, violet, and occasionally orange and white. Typically, when the particles collide with oxygen, yellow and green are produced. Interactions with nitrogen produce red, violet, and occasionally blue colors.

The type of collision also makes a difference to the colors that appear in the sky: atomic nitrogen causes blue displays, while molecular nitrogen results in purple. The colors are also affected by altitude. The green lights typically in areas appear up to 150 miles (241 km) high, red above 150 miles; blue usually appears at up to 60 miles (96.5 km); and purple and violet above 60 miles.

These lights may manifest as a static band of light, or, when the solar flares are particularly strong, as a dancing curtain of ever-changing color.

Credit: Thinkstock

These astonishing light shows are the result of charged particles interacting with Earth's magnetic field

WHERE AND WHEN TO SEE THE LIGHTS

The best places to see the northern lights are Alaska and northern Canada, but visiting these vast, open expanses is not always easy. Norway, Sweden and Finland also offer excellent vantage points. During periods of particularly active solar flares, the lights can be seen as far south as the top of Scotland and even northern England. On rare occasions, the lights can be seen farther south.

The northern lights are always present, but winter is usually the best time to see them, due to lower levels of light pollution and the clear, crisp air. September, October, March and April are some of the best months to view the aurora borealis. The lights are known to be brighter and more active for up to two days after sunspot activity is at its highest. Several agencies, such as NASA and the National Oceanic and Atmospheric Administration, also monitor solar activity and issue aurora alerts when they are expected to put on a particularly impressive show.

IS EARTH'S MAGNETIC FIELD FLIPPING SOON?

EARTH'S NORTH MAGNETIC POLE IS SO OUT OF WHACK THAT SCIENTISTS NEED TO UPDATE THE GLOBAL MAGNETIC-FIELD MODEL THEY RELEASED JUST A FEW YEARS AGO. COULD THAT BE A SIGN THAT THE MAGNETIC POLES WILL FLIP SOON?

WORDS: ELIZABETH HOWELL

The World Magnetic Model (WMM) is the name of the updated representation of the magnetic field of Earth. The magnetic pole is moving erratically out of the Canadian Arctic and toward Siberia so unpredictably that it has taken scientists by surprise. That 2015 update was supposed to remain valid until 2020, Arnaud Chulliat, a geomagnetist at the University of Colorado Boulder and the National Oceanic and Atmospheric Administration's (NOAA) National Centers for Environmental Information, told Nature.

It's no news that the pole is moving; long-term records from London and Paris (kept since 1580) show that the north magnetic pole moves erratically around the rotational north pole over periods of a few hundred years or longer, Ciaran Beggan, a geophysicist with the British Geological Survey who is involved in WMM updates, told Space.com. He cited a 1981 study from the journal Philosophical Transactions of the Royal Society of London.

But what's really catching attention is the acceleration in movement. Around the mid-1990s, the pole suddenly sped up its movements from just over 9 miles (15 kilometers) a year to 34 miles (55 kilometers) annually.

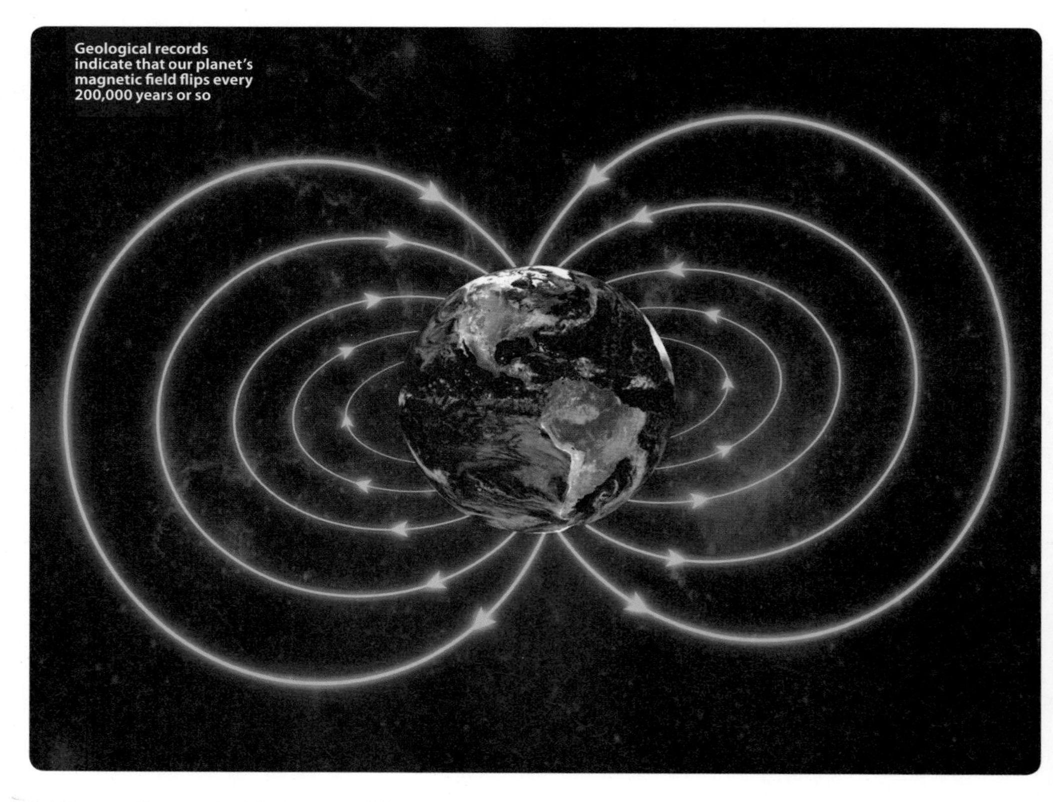

Geological records indicate that our planet's magnetic field flips every 200,000 years or so

As of last year, the pole careened over the international date line toward the Eastern Hemisphere. The chief cause of the movement comes from the Earth's liquid-iron outer core, which is also called the "core field." Smaller factors also affect the movement.

Those influences include magnetic minerals in the crust and upper mantle (especially for local magnetic fields) and electric currents created by seawater moving through an "ambient magnetic field," according to the 2015 report of the WMM. "One of the reasons we can update the map is that the European Space Agency launched a set of highly accurate magnetic-field satellites in 2013," said Beggan, referring to Swarm. "We have a superb data set from which we can make very good magnetic-field maps and update them every six to 12 months," Beggan added. "We noticed that the specification of the WMM was not being met in the high-latitude region around the pole, as the error exceeded 1 degree of grid angle on average. This triggered us to examine whether it was worth issuing a new update." The interest in this phenomena isn't just academic, either. Throughout Earth's history, the north and south magnetic poles have regularly flipped. Is that about to happen again?

Our planet's magnetic field is like a protective shield against some of the sun's intense radiation

The aurora is the result of interactions between Earth's magnetic field and charged particles from the sun

"THE EFFECTS OF A MAGNETIC FIELD FLIP COULD BE PROFOUND ON TECHNOLOGY"

FLIPPING OUT

What's more, the core field appears to be weakening – which may be a sign that the planet's magnetic field will flip. Earth's north and south poles periodically swap locations, with the last flip happening about 780,000 years ago. The poles also weakened temporarily and rapidly about 41,000 years ago, Beggan added, but never underwent a full flip. A 2018 study in the journal Proceedings of the National Academy of Sciences suggested that Earth's magnetic field got substantially weaker before the last big changeover.

While any magnetic-field flip would still be thousands of years away, the effects could be profound on technology, if similar to today's technology. This is because the weaker magnetic field would be somewhat poorer at shielding Earth against the solar wind (the constant stream of charged particles emanating from the sun) and cosmic rays (blasts of radiation from deep space). Magnetic compasses would not be as accurate, and satellites that monitor the weather or carry telecommunications signals could be disrupted, said Monika Korte, head of GFZ Potsdam's working group on geomagnetic field evolution in Germany.

"Regarding increased radiation, that would go along with decreased shielding, [but] it seems that the atmosphere would still provide sufficient shielding at Earth's surface that humans and animals would not be significantly affected," Korte told Space.com. "However, all the effects we currently only see during strong solar/geomagnetic storms would likely increase and occur [...] during moderate solar activity," she added. "This includes satellite outages or damage to satellites, increased radiation doses on long-distance aircraft and the ISS [the International Space Station], [and] distortions of telecommunication and GPS signals." Ongoing monitoring of the Earth's geomagnetic field will continue after the WMM release, principally through the European Space Agency Swarm mission, Korte said. But she noted that measuring the north magnetic pole's position is a challenge. That's because the pole is located in a remote area and the measurement of Earth's magnetic field is influenced by all magnetic-field sources – including the magnetic fields found in Earth's atmosphere (the ionosphere and magnetosphere).

"It will depend on the future magnetic field change, which we cannot predict, if another update to the model out of the usual schedule might be required," she added.

HOW DID PLUTO GET ITS 'WHALE'?

WHAT THE DATA FROM NEW HORIZONS CAN REVEAL ABOUT THIS DISTANT DWARF PLANET

WORDS: MEGAN GANNON

In 2015, scientists learned that there's a giant red "whale" on Pluto. This dark-colored region could be the mark of a giant impact – the same one that produced Pluto's huge moon Charon, according to a group of researchers in Japan.

The surface of Pluto – the biggest object inside the Kuiper Belt (the ring of ice bodies beyond Neptune's orbit) – remained mysterious for decades. Astronomers knew the dwarf planet as little more than a blurry orb until NASA's New Horizons probe revealed its surprisingly complex features in high definition during a flyby in July 2015. Thanks to that mission, we now know that Pluto has towering ice mountains, blue skies, a 620-mile-wide (1,000-kilometer-wide) heart-shaped nitrogen glacier, jagged faults and, potentially, a sub-surface ocean.

One of the most prominent features on Pluto is the informally named Cthulhu Regio, also known as "the whale," which stretches across 1,900 miles (3,000 kilometres). Cthulhu Regio is pockmarked with craters, which suggests it is billions of years old – much older than the craterless, young "heart" it borders. Scientists have said the dark region's reddish coloring might come from tholins, which are complex hydrocarbons.

To further investigate how the whale got its coloring, Yasuhito Sekine, an associate professor at the University of Tokyo, conducted heating experiments on organic molecules, like formaldehyde, that would have been present on the newly forming Pluto soon after the formation of the solar system. Sekine found that he could produce the same dark, reddish color after heating solutions above 122 degrees Fahrenheit (50 degrees Celsius) for several months.

Meanwhile, Hidenori Genda, an associate professor at the Tokyo Institute of Technology, conducted computer simulations of a giant impact on Pluto. Genda found that the same impact that would have created Pluto's moon Charon, which is about half the size of Pluto, could have created a huge pool of hot water near Pluto's equator. And as this giant hot-water pool cooled, reddish complex organic materials would have formed, according to the findings, which were published in 2017.

"For the terrestrial planets in our solar system, giant impacts are common," Genda explained to Space.com. "Our results suggest that giant impacts are common in [the] outer system beyond the Neptune orbit."

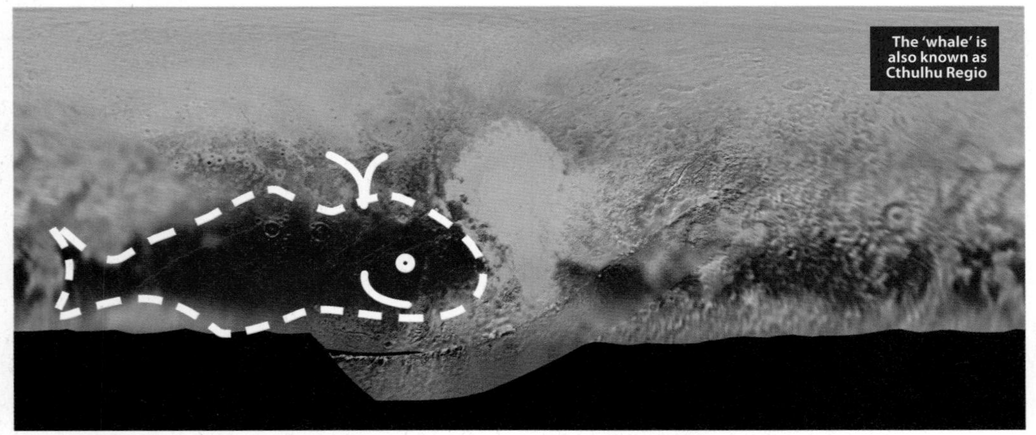

The 'whale' is also known as Cthulhu Regio

Credit: NASA/Johns Hopkins University Applied Physics Laboratory/Southwest Research Institute

THE COLOR OF AN IMPACT

The color changes across Pluto "make an interesting pattern, and we don't have good ideas to explain all of these features, so we are all still in the initial stages [of] exploring different hypotheses to explain these variations," said Kelsi Singer, a co-investigator on the New Horizons extended mission at the Southwest Research Institute in Colorado.

But Singer, who wasn't involved in the new study, isn't entirely convinced by the newly described impact scenario. She told Space.com in an email that it is unlikely that the "whale" has remained mostly the same for the past 4 billion years or so, because the region has a lot of variation within it.

"There are some heavily cratered areas, and some smoother, almost crater-free areas that have younger ages," Singer said. "You could perhaps argue that if the dark layer was very thick (more than a few km) you could keep it around for 4 billion years or longer and still have other craters form in it, tectonics fracturing it, and keep it dark."

But Singer said that in most areas of the whale, the dark material does not seem to be thick; thinner patches of the dark material are sitting on top of a brighter surface. Singer thinks a simpler explanation for the whale's coloring could be that the dark material formed from methane being processed by radiation on Pluto's surface or in the atmosphere. To fully confirm the chemical makeup and impact history of Pluto, scientists might need to send additional space telescopes to observe the dwarf planet, or perhaps, one day, a probe to sample the icy surface.

"If we can get some information on chemical compositions of complex organic matters in the whale region, it will help us to confirm or deny the impact origin of this region," Genda said. "UV spectra will give us that information, but unfortunately, New Horizons did not have a UV spectra instrument. Ultimately, sample return from the whale region can reveal the origin of this region."

Credit: NASA/JHUAPL/SwRI

Above: In this image, the head of Pluto's "whale" is visible on the lower left

Credit: NASA/Johns Hopkins University Applied Physics Laboratory/Southwest Research Institute/Lunar and Planetary Institute

New Horizons' close-up views have revolutionized our understanding of Pluto

EXPLORATION

Credit: NASA

Credit: NASA, ESA/Hubble and the Hubble Heritage Team

Credit: NASA,

118
Credit: NASA

120
Credit: NASA/JPL-Caltech

126

СОЮЗ
136
K. RUBINS
K. РУБИНС
S. RYZHIKOV
S. KUD-SVERCHKOV
РОССИЯ

MOST EXTREME HUMAN SPACEFLIGHT RECORDS

A LOOK AT SOME OF HUMANITY'S RECORD-SETTING SPACEFLIGHT ACHIEVEMENTS

WORDS: MIKE WALL

On April 12, 1961, humanity became a spacefaring species when cosmonaut Yuri Gagarin blasted into orbit on a 108-minute flight high above Earth. So Gagarin set the original record – first person in space. But over the years, people have notched many other records as our species has extended its toehold in the cold depths of space. Here's a look at some of these marks, from the oldest person in space to the most consecutive days spent away from terra firma.

Cosmonaut Yuri Gagarin in the Vostok capsule that took him into space in 1961

FIRST PEOPLE IN SPACE

Gagarin was the first person to fly in space, and the first American followed only a few weeks later. Alan Shepard blasted off on Freedom 7 on May 5, 1961. The first female in space was Valentina Tereshkova, a Russian cosmonaut, who flew in space in June 1963. There were several other female cosmonauts selected with her, but none of the others flew. The next woman in space, Svetlana Savitskaya, didn't fly until 1982. The first American woman in space was Sally Ride, who reached space on June 18, 1983 as part of the space shuttle mission STS-7.

For almost 20 years, the Americans and the Soviets were the only nations with astronauts. The first nation outside of those two countries to fly an astronaut was the former Czechoslovakia, which saw Vladimir Remek fly on the Soviet Soyuz 28 mission in 1978. Since then, dozens of nations from all over the world have seen their citizens fly in space on American, Soviet or Russian spacecraft.

Bruce McCandless performing an untethered spacewalk with the Manned Maneuvering Unit

FIRST SPACEWALKS

The first-ever spacewalk was performed by Alexei Leonov, who performed a 12-minute spacewalk during Voshkod 2 on March 18, 1965. The astronaut later said that he had trouble getting back inside the spacecraft (his spacesuit ballooned) and that he also was close to getting heatstroke, but he made it back home safely.

The first American spacewalk was performed by Ed White on June 3, 1965. The first spacewalk by a woman wasn't for nearly 20 years afterwards, when Svetlana Savitskaya performed a spacewalk on July 25, 1984, outside of the Salyut 7 space station. The first American woman to perform a spacewalk was Kathryn Sullivan, who left space shuttle Challenger on Oct. 11, 1984. The first untethered spacewalk (one of only a handful performed) happened on Feb. 7, 1984, when Bruce McCandless used the Manned Maneuvering Unit to move away from space shuttle Challenger during mission STS-41-B.

John Glenn's portrait for the STS-95 Space Shuttle mission

OLDEST PERSON IN SPACE

U.S. Sen. John Glenn, D-Ohio, was 77 when he flew on space shuttle Discovery's STS-95 mission in October 1998. The mission marked Glenn's second spaceflight; he had become the first American to orbit the Earth back in February 1962. So Glenn holds another record as well: the longest time between trips to space (36 years 8 months). The oldest woman in space was Peggy Whitson, who was 57 years old during her last flight (Expeditions 50, 51 and 52 in 2016-2017).

Gherman Titov was just 25 when he orbited Earth in 1961

YOUNGEST PERSON IN SPACE

Cosmonaut Gherman Titov was one month shy of his 26th birthday when he launched into orbit aboard the Soviet spacecraft Vostok 2 in August 1961. He was the second person to orbit the Earth, performing 17 loops around our planet during his 25-hour flight. Titov was also the first person to sleep in space, and reportedly the first to suffer from "space sickness" (motion sickness in space). Tereshkova was not only the first woman in space, but she still hold the record as the youngest, at 26.

Gennady Padalka has spent over two years in space during his cosmonaut career

MOST TOTAL TIME SPENT IN SPACE

Cosmonaut Gennady Padalka holds this record, with a little more than 878 days in space accrued over five spaceflights. That's almost two and a half years (2 years, 4 months, 3 weeks, and 5 days) spent zipping around the Earth at about 17,500 miles per hour (28,164 kilometres per hour). For women, the record is held by NASA astronaut Peggy Whitson, who spent more than 665 days in space. That also happens to be the endurance record for any American astronaut.

MOST EXPENSIVE SPACESHIP

Member nations began building the International Space Station – which is about as long as a football field and boasts as much living space as a five-bedroom house – back in 1998. It was completed in 2012, although more expansions are in store.

The cost for the orbiting lab was estimated at $100 billion in 2011. That makes the station the single most expensive structure ever built. The cost will continue to rise due to more modules and time operating the station.

LARGEST SPACESHIP EVER BUILT

Once again, the International Space Station is the winner. The orbiting lab is the product of five space agencies representing more than 15 countries. From one end of its backbone-like main truss to the other, it measures about 357.5 feet (109 meters) across. There are huge solar arrays at each end of the truss, and they have a wingspan of 239.4 feet (73 m).

Astronauts live inside a series of connected, pressurized modules that are attached to the main truss. These modules have a habitable space roughly equivalent to the interior cabin of a Boeing 747 jumbo jet. The station is currently staffed by six astronauts, but that population has jumped to between nine and 13 people when a visiting vehicle was docked.

The space station is so large that it can easily be seen by the unaided eye from the ground if skywatchers have clear skies and know where to look. The station appears as a fast-moving bright light that can outshine the brightest star (Sirius) or Venus, depending on the viewing conditions where you are.

NASA are due to keep funding the ISS until at least 2025

BIGGEST SPACE GATHERING

It may sound unlucky, but the record for the largest human gathering in space stands at 13 – which was set during NASA's STS-127 shuttle mission aboard Endeavour in 2009. In July 2009, Endeavour docked with the International Space Station. The shuttle's seven-person crew then went aboard the orbiting lab, joining the six spaceflyers already there. The 13-person party was the largest-ever gathering of people in space at the same time. While subsequent NASA shuttle and station crews matched the 13-person record, it has never been topped.

The astronauts and cosmonauts aboard the ISS during STS-127 set the record for the most humans in space at the same time

Clockwise from top left: Metcalf-Lindenburger, Yamazaki, Wilson and Dyson

MOST WOMEN IN SPACE AT ONCE

This record stands at four women in orbit at the same time. In April 2010, NASA astronaut Tracy Caldwell Dyson traveled to the International Space Station in a Russian Soyuz spaceship. She was soon joined on the orbiting lab by NASA astronauts Stephanie Wilson and Dorothy Metcalf-Lindenburger and Japan's Naoko Yamazaki, who made the trip aboard the space shuttle Discovery on its STS-131 mission.

Cosmonaut Anatoly Solovyev continues to hold the record for the most spacewalks

MOST SPACEWALKS

Russian cosmonaut Anatoly Solovyev made 16 spacewalks over the course of five missions in the 1980s and 1990s. Solovyev spent more than 82 hours outside his spacecraft on those excursions – another record.

NASA astronaut Michael Lopez-Alegria holds an American record of 10 spacewalks, with a total time spent outside of 67 hours 40 minutes. Close behind is the woman to make the most spacewalks; American astronaut Peggy Whitson, performed 10 spacewalks over multiple missions for a total time of 60 hours 21 minutes.

LONGEST SINGLE SPACEWALK

On March 11, 2001, NASA astronauts Jim Voss and Susan Helms spent 8 hours and 56 minutes outside the space shuttle Discovery and the International Space Station during the STS-102 mission, performing some maintenance work and preparing the orbiting lab for the arrival of another module. It remains the longest spacewalk in history.

Helms pictured during the record-breaking spacewalk

During their return to Earth, the Apollo 10 crew travelled faster than any other humans ever had or have since

FASTEST HUMAN SPACEFLIGHT

The crew of NASA's Apollo 10 moon mission reached a top speed of 24,791 mph (39,897 kph) relative to Earth as they rocketed back to our planet on May 26, 1969. That's the fastest any human beings have ever traveled. The Apollo 10 mission served as a dress rehearsal for NASA's first moon landing two months later, on July 20, 1969. Apollo 10 astronauts Cernan, John Young and Tom Stafford orbited the moon in their command module Charlie Brown and Lunar Module Snoopy. Later, Stafford and Cernan took the Snoopy lunar lander down to within 50,000 feet (15,243 meters) of the moon's surface before returning to dock with the Charlie Brown module.

Space Shuttle Columbia's intended landing date was postponed due to bad weather

LONGEST SPACE SHUTTLE MISSION

The space shuttle Columbia launched on its STS-80 mission on November 19, 1996. It was originally slated to return to Earth that December 5, but bad weather pushed the landing back by two days. When Columbia finally came home, it had spent nearly 17 days and 16 hours in space – a record for a space shuttle mission.

MOST TIME ON THE MOON

In December 1972, Harrison Schmitt and Eugene Cernan of NASA's Apollo 17 mission spent just under 75 hours – more than three days – poking around on the surface of the moon. They also performed three moonwalks that lasted a total of more than 22 hours. Perhaps the astronauts lingered because they suspected humanity wouldn't be back for a while – Apollo 17 marked the last time people traveled to the moon, or even went beyond low-Earth orbit.

Cernan (left) and Schmitt (right) aboard the Apollo 17 spacecraft

Valery Polyakov seen here peering through the Mir Space Station as the Space Shuttle Discovery is about to dock

FARTHEST AWAY

The record for the greatest distance from Earth has stood for more than four decades. In April 1970, the crew of NASA's Apollo 13 mission swung around the far side of the moon at an altitude of 158 miles (254 km), putting them 248,655 miles (400,171 km) away from Earth. It's the farthest our species has ever been from our home planet.

A photo of Earth taken during the Apollo 13 mission which narrowly avoided disaster

MOST CONSECUTIVE DAYS IN SPACE

Russian cosmonaut Valery Polyakov spent nearly 438 consecutive days aboard the Mir space station, from January 1994 to March 1995. He therefore holds the record for longest single human spaceflight – and perhaps set another one for wobbliest legs when he finally touched down. The most consecutive days in space by an American is 340 days, which happened when Scott Kelly took part in a one-year mission to the International Space Station in 2015-16 (along with cosmonaut Mikhail Kornienko). The longest single flight by a woman took place in 2016-17, when NASA's Peggy Whitson spent 288 days aboard the space station. NASA astronaut Christina Koch is due to break that record when she returns from a 328-day mission to the space station in spring of 2020.

The first ISS crew were (left to right) Yuri Gidzenko, William Shepherd and Sergei Krikalev

LONGEST CONTINUOUSLY INHABITED SPACECRAFT

This record belongs to the International Space Station, and it grows every day. The $100 billion orbiting lab has been continuously occupied since Nov. 2, 2000. This span of time – plus two days, since the first station crew launched Oct. 31, 2000 – also marks the longest period of continuous human presence in space.

Chang-Diaz (left) and Ross (right) share the record of most spaceflights

MOST SPACEFLIGHTS BY AN ASTRONAUT

This one is shared by two NASA astronauts. Franklin Chang-Diaz and Jerry Ross both went to space seven times aboard NASA's space shuttles. Chang-Diaz made his flights between 1986 and 2002, while Ross made his between 1985 and 2002.

THE HUBBLE SPACE TELESCOPE

DISCOVER THE ICONIC INSTRUMENT WITH ONE OF THE GREATEST REDEMPTION STORIES IN SCIENCE HISTORY

WORDS: CALLA COFIELD

Throughout its time in space, the iconic Hubble Space Telescope – which launched on April 24, 1990 – has provided spectacular views of the cosmos and revealed exceptional insights about the universe. But there were also moments when it looked as though decades of work, plus billions of taxpayer dollars, might suddenly slip down the drain, and there were worries that the project might fail completely.

But Hubble overcame those obstacles to become one of the most successful telescopes ever built, both in terms of its scientific return and its impact on the public. And after more than 25 years of operation, Hubble's best days may still be ahead of it, astronomers say.

"Even the most optimistic person to whom you could have spoken back in 1990 couldn't have predicted the degree to which Hubble would rewrite our astrophysics and planetary science textbooks," former NASA Administrator Charles Bolden said at an event in 2015 to celebrate the telescope's 25th anniversary. "A quarter-century later, Hubble has fundamentally changed our understanding of our universe, and our place in it."

At its current pace, the Hubble telescope produces 10 terabytes of new data per year – enough to fill the entire collection of the Library of Congress, Bolden said. At that same event, Kathy Flanagan, interim director of the Space Telescope Science Institute in Baltimore, which operates Hubble's science program, said scientists using data from the telescope have produced "nearly 13,000" science papers.

DID YOU KNOW...?

Hubble has made over 1.3 million observations since 1990 when its mission began

THE CLIMB TO SUCCESS

The Hubble telescope climbed to its current position at the peak of accomplishment from some deep valleys of near-failure. In his book, "The Universe a Mirror: The Saga of the Hubble Space Telescope and the Visionaries Who Built it," (Princeton University Press, 2008), science writer Robert Zimmerman chronicled the decades-long slog to get the Hubble telescope to where it is today. First, there was the chore of convincing the astronomy community to agree to invest in such a costly project, and then to get Congress to fund it, and to keep funding it during construction.

It wasn't just the telescope that suffered during those years; Zimmerman also wrote about people who dedicated themselves to Hubble at the expense of their careers or even their personal lives.

The Hubble Space Telescope was originally scheduled to blast off in 1983 but didn't get off the ground until 1990. Shortly after the telescope's launch, the scientific team realized the images they were receiving were blurry. It turned out that the telescope's mirror was ground ever so slightly to the wrong thickness. (The flaw arose because of a mistake with the testing equipment used during the mirror's construction.)

In 1993, the first Hubble servicing mission installed hardware that could adjust for the flaw in the mirror, and the telescope quickly blossomed to its full potential. It revealed new information at

Credit: ESA

Hubble has captured some of the most stunning images of the cosmos, including this view of the Antennae Galaxies

"NASA HAS NO FIRM DECOMMISSIONING DATE BECAUSE THE OBSERVATORY IS OPERATING BETTER THAN ANYONE EXPECTED"

every size scale, from the solar system to the entire observable universe. Hubble has found four new moons around Pluto, demonstrated that galaxies frequently collide and merge together, drastically improved measurements of the age of the universe, and showed that space is not only expanding but spreading out faster and faster.

By 2003, Hubble had provided more than a decade of valuable science and beautiful images. At that point, it could have retired and still been labeled a success. But plans were in the works to add two new instruments to Hubble and repair two instruments that had stopped working.

The fifth and thus-far final crewed repair mission to Hubble took place in 2009. That mission is a microcosm of Hubble's life story: full of close calls that nearly spelled disaster for the telescope, like when a bolt holding down a handrail wouldn't come loose and nearly prevented the astronauts from getting to one of the instruments that

Hubble's Wide Field Camera 3 captured this image of Stephan's Quintet

needed fixing. In the end, the mission was a complete success. The astronauts installed two new instruments, fixed two broken instruments, and installed new batteries, new gyroscopes and a new scientific computer, to prolong Hubble's life. Today, it continues to be one of the most powerful, most in-demand telescopes in the world.

WHAT THE FUTURE HOLDS

So, what's next for the Hubble Space Telescope? "Frankly, we never even thought that the telescope would last this long," Bolden said at the image-unveiling event in 2015. "The original plan for Hubble, we were told, was maybe 15 years. The fact that we are still going strong a quarter-century later is thanks to the Hubble heroes [...] many of whom you will never know."

Hubble will stop taking data someday, but right now, NASA has no firm decommissioning date because the observatory is operating better than anyone expected, even more than five years after its last servicing.

Right now, the Hubble team members aim to keep the telescope running through at least June 2021. If Hubble can reach that goal, it should overlap with NASA's James Webb Space Telescope, which is set to launch into space in March 2021.

While the Hubble telescope sees mostly optical and ultraviolet light, the $8.8-billion James Webb Space Telescope sees infrared light, and it will peer even deeper into the universe than Hubble has. The James Webb telescope has a larger mirror – 21.3 feet (6.5 meters) wide, compared to Hubble's 7.9-foot (2.4-metre) mirror – and will have a more powerful camera. And yet, it's hard to think how any future telescopes will fill the shoes Hubble leaves behind.

The Hubble telescope cannot maintain its orbit forever – if left alone, it will fall to Earth and be destroyed, likely in the mid- to late 2030s. NASA officials have said they won't let an uncontrolled re-entry happen, because people on the ground could be hurt by falling Hubble parts. So the agency has two options: Either steer Hubble to a safe destruction over the Pacific Ocean, or boost the telescope to a higher orbit (and possibly refurbish it one more time).

The time frame of Hubble's ultimate fate remains up in the air, because no one knows for sure how much longer Hubble will keep producing good science. Zimmerman said he'll bet that if the Hubble telescope is still working when the time comes to capture it, NASA will find a way to put it back into a steady orbit.

Many years after the Hubble Space Telescope's deployment, the iconic observatory's birthday celebrations are by no means a memorial. Hubble is currently performing better than when its mission started, and shows no signs of slowing down. In fact, the telescope that has ascended to such great heights may not yet have reached the pinnacle of its astonishing accomplishments.

Hubble's crystal-clear view of the Sombrero Galaxy, 30 million light years away

Credit: NASA/ESA and The Hubble Heritage Team (STScI/AURA)

This shot of the star cluster Westerlund 2 was chosen as Hubble's 25th anniversary image

Credit: NASA, ESA, the Hubble Heritage Team (STScI/AURA), A. Nota (ESA/STScI), and the Westerlund 2 Science Team

One of Hubble's most iconic images is the 'Pillars of Creation' in the Eagle nebula

Credit: NASA, ESA/Hubble and the Hubble Heritage Team

THE INTERNATIONAL SPACE STATION

HOW THE SPACE LAB HAS EVOLVED DURING ITS TWO DECADES IN ORBIT

WORDS: DORIS ELIN SALAZAR

The station pictured in 2010 by the departing crew of STS-132

For two minutes, NASA astronaut Leroy Chiao could see nothing but the blue-marbled Earth swirling above his head. "Surreal" is how he described the moment. Chiao was in the middle of a lengthy spacewalk to assemble part of the International Space Station, an ambitious orbiting laboratory the likes of which had never been seen before.

November 20th 2018 marked the 20th anniversary of the launch of the International Space Station's first component, made possible by the contribution of hundreds of engineers, space shuttle astronauts like Chiao, international support and the crews who continue launching up to this day. The station itself has been continuously occupied since Nov. 2nd 2000.

As we reflect on what the space station has afforded humanity – diplomacy, progress in human spaceflight and discoveries in the life sciences – we also have to wonder what the space laboratory's future holds.

The International Space Station, or ISS, pushed NASA into "an entirely new way of thinking," Gary Oleson, a station engineer, told Space.com. Oleson worked as a member of NASA's Space Station Program Office from 1988 to 1993; at first doing the cost side of systems engineering, and then as a principal systems engineer liaison, where he focused on the project's logistics and maintenance.

"We normally think of a spacecraft as being a spacecrawvft," Oleson said. "But it turned out that the International Space Station, during the assembly, was not, from an engineering point of view, one spacecraft. It was at one time 19 different spacecraft, because every time you went up and added a new element, you had a different spacecraft. It had a different mass; it had a different reliability."

That's just one reason why the team was thrilled that space station construction went so smoothly. "We were kind of surprised when we didn't have bigger technical hiccups during the assembly phase," Chiao told Space.com. "The pieces actually all fit together. Pieces that were built in other countries and using different electric systems… and a lot of these pieces weren't even fit-checked."

The ISS includes contributions from 15 nations. The space lab's major partners include

Credit: NASA

DID YOU KNOW...?

As of March 2021, over 240 people from 19 countries have visited the International Space Station

The station's Cupola module's seven windows provide astronauts with stunning views of Earth below

Credit: NASA JSC

NASA, Russian space agency Roscosmos and the European Space Agency; the Japan Aerospace Exploration Agency and the Canadian Space Agency are also partners. Since the space station opened, the ISS has hosted 232 individuals from 18 countries.

The space lab flies at an average altitude of 248 miles (400 kilometers) above Earth's surface. It circles the planet every 90 minutes at a speed of about 17,500 mph (28,000 km/h). To put that in perspective, the distance the ISS travels daily is roughly how far it would take to go from Earth to the moon and back.

Current plans call for space station operations to continue through 2024. The Trump administration has proposed to no longer directly support the space station after 2025.

"[NASA] intends to be a major customer for whoever operates the station, but it would like to get into the role of being the primary payer for the operation of the station," John Logsdon, Professor Emeritus at the Space Policy Institute in George Washington University, told Space.com. "I think the crucial issue is how long the government pays for the operation of the station and is there some sort of private operator to step in and take over the station.

Chiao said he's disappointed in the plan to discontinue the space station after 2024 or 2025. "This whole idea of commercializing it… the space station was never designed or intended to be profitable," he said. "Just to pay for that infrastructure, it's unreasonable. And launch costs. How are you going to cover launch costs for research and astronauts to and from the space station commercially? It just doesn't make any sense. If we keep going down this path, the space station will end."

Technologically, of course, the space station must end someday – and when the project began it was designed to last only 15 years. But current evaluations suggest that most of what's up there now should be perfectly safe through 2028, if not beyond.

The argument to end the station's public-service career early is that saving on the costs of the space station means funding more human space exploration like a return to the Moon and an extension to Mars.

"I think the lunar base makes sense for a lot of reasons," Chiao said. "You want to be able to go, reestablish your environment and make sure it's going to work before you send it off to Mars." But, he added, "saving in one area doesn't mean funds are going to be available for the program you want… that's not necessarily true."

While the future of the space station remains uncertain, the perspective it affords to anyone looking at images taken from its cupula is most definitely awe-inspiring.

"The fact that we do have a vehicle in space – the ISS – and the fact that we do launch regularly to it, serves as an inspiration to young people," Chiao added. Stories like his breathtaking two minutes gazing at the continents and clouds swirling across the planet are enough to mesmerize anyone.

"WHEN THE PROJECT BEGAN IT WAS DESIGNED TO LAST ONLY 15 YEARS. BUT MOST OF IT SHOULD BE PERFECTLY SAFE THROUGH 2028"

Credit: NASA.

Astronauts regularly conduct spacewalks to perform maintenance on the station

Credit: NASA JSC

Canadarm 2, the station's robotic arm, is used to assist with station assembly, docking and maintenance

SPACE STATION STATS

HERE'S A LOOK AT THE INTERNATIONAL SPACE STATION IN NUMBERS

WORDS: REMY MELINA

$100 BILLION

Estimated cost of the ISS. This gives the space station the grandiose title of being the world's most expensive single object

925,000

How much the ISS weighs in pounds (419,600 kilograms), the equivalent of more than 320 cars

357

Overall length of the station in feet (109 meters). It's about the length of a U.S. football field, including its backbone-like truss segments and solar wings

135

How many times you would have to cross North America to travel a distance equal to that traveled by the ISS in one day (about the distance to the moon and back)

Tito (left) with cosmonauts Talgat Musabayev (center) and Yury Baturin (right)

Credit: NASA JSC

$20 MILLION

The amount that Dennis Tito, an American multimillionaire entrepreneur, paid to become the first person ever to fly to the station as a self-funded space tourist. He stayed on the station for eight days before flying back to Earth

75–90

The number of kilowatts of power that is supplied by an acre of solar panels

3

The size of the station's first crew in 2000, and the number of astronauts or cosmonauts ferried to the station on a single Russian Soyuz spacecraft

Credit: NASA

340 The number of days that astronaut Scott Kelly spent on the station. He holds the record for the longest single mission

90 The number of minutes it takes the ISS to circle the Earth as it travels at 4.8 miles (7.7 km) per second

52 The number of computers aboard the ISS to control its systems

8 The total length, in miles, of wire that connects the electrical power system (12.9 km)

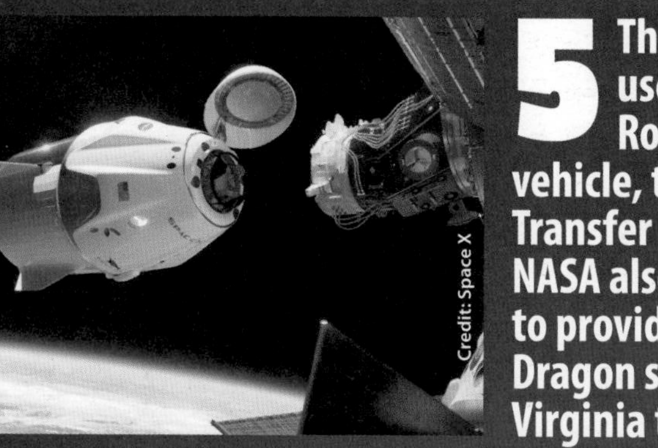

Credit: Space X

5 The number of unmanned spacecraft used to haul supplies to the space station. Robotic spacecraft include Russia's Progress vehicle, the European Space Agency's Automated Transfer Vehicle and Japan's H-2 Transfer Vehicle. NASA also has contracts with the company SpaceX to provide cargo flights using its unmanned Dragon spacecraft, and with Orbital Sciences in Virginia to do the same on its Cygnus spacecraft

4 The tons of food required to support a crew of three for about six months (3,630 kg). The ISS crews' favorites include shrimp cocktail, tortillas, barbecue beef brisket, breakfast sausage links, chicken fajitas, vegetable quiche, macaroni and cheese, candy-coated chocolates and cherry blueberry cobbler. Lemonade is the most popular drink

5 Number of space agencies that contributed to building the station. NASA, Russia's Roscosmos, the Japanese space agency, the Canadian Space Agency and the European Space Agency all contributed to the construction of the station

6 The number of months that an astronaut typically lives and works on the ISS during a mission

Data sourced from NASA, correct as of August 2017

EXPLORATION

A YEAR IN SPACE

TWO MEN SPENT 340 DAYS IN SPACE. SCIENTISTS ARE STILL FIGURING OUT WHAT THEY'VE LEARNED

WORDS: MEGHAN BARTELS

Between March 2015 and March 2016, a man spent a total of 357 hours being poked and prodded: sacrificing blood and urine, sweating through sprints and letting others read his journals – oh, and living on the International Space Station.

The man was Scott Kelly, and with his Russian counterpart, Mikhail Kornienko, he spent a total of 340 days living in space as part of NASA's first attempt to understand how such long stints without gravity affect the human body. But while the men returned to Earth nearly three years ago, scientists are still trying to figure out what they've learned. A new paper lays out the tests built into the "One-Year Mission" of Kelly and Kornienko and contextualizes it in the context of longer spaceflight – like that which would required by a mission to Mars, lasting at least two years.

What stands out most is how little NASA knows about how the body might respond to such conditions: Before Kelly's launch, no U.S. astronaut had ever spent more than six consecutive months in space, so the agency was desperate for data. (Russia had sent six cosmonauts to live on the Mir space station for 300 days or more, but doesn't make that data public.)

So before the flight, scientists came up with a suite of 17 different investigations the pair of astronauts would participate in – tackling unknowns like how fluids move through the body in microgravity, how they slept and how the communities of microbes living on and inside them changed. By running tests on the two astronauts before, during and after their 11-month stay, they could see how the astronauts' bodies responded to a long spaceflight. They could also compare the measurement with data gathered in conjunction with previous, shorter spaceflights. And in a separate analysis, Kelly's data is also being compared to data about his twin brother, NASA astronaut Mark Kelly, who stayed on Earth as a control.

But even the One-Year Mission, the best study scientists have to date on the effects of long-term spaceflight only includes two test studies – and those are both men, both Caucasian, both in their early-to-mid 50s. And the 11-month flight doesn't match the timeline needed to get to Mars. However, the 2019 paper argues that the study acts as a crucial foundation for future, larger studies, and that comparing six-month and one-year results will help scientists better extrapolate the results out to longer flights.

Credit: NASA

Knowing the health effects longer space missions will be crucial in getting humans to Mars

Credit: NASA/Bill Stafford

Studies on Kelly (left) and Kornienko (right) will help scientists understand how long-duration space flights may affect the human body

The next step, the researchers argue, is to build up a larger study. That would use the same tests and procedures developed for Kelly and Kornienko, but apply them more broadly: to 10 astronauts on year-long missions, 10 on six-month missions and 10 on two-month missions. Doing so, the authors write, will help space agencies further bridge the gaps in research; NASA has already expressed interest in proposals along these lines.

The overview of the research was included in a paper published Jan. 1 2019 in the journal Aerospace Medicine and Human Performance. The results of individual studies conducted as part of the One-Year Mission will be published separately.

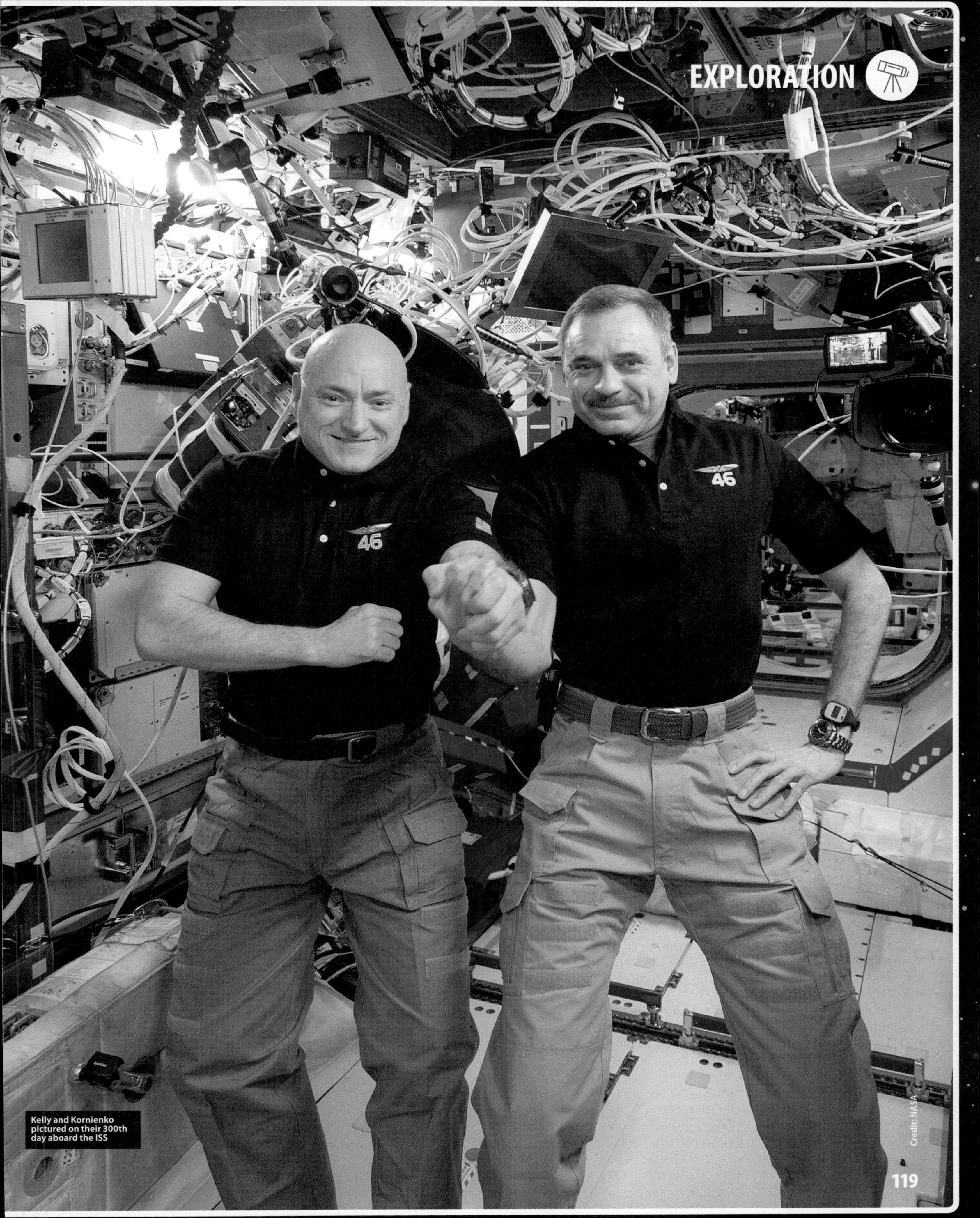

Kelly and Kornienko pictured on their 300th day aboard the ISS

Credit: NASA

BIZZARE MARS

THE WEIRDEST MARTIAN DISCOVERIES BY THE OPPORTUNITY AND SPIRIT ROVERS

WORDS: HANNEKE WEITERING

NASA's Opportunity rover has officially concluded its 15-year mission on Mars, the agency announced in February 2019. Eight months after a raging dust storm incapacitated Opportunity's solar panels, leaving it unable to communicate with Earth, NASA has stopped waiting for it to wake up and will no longer listen for signals from the rover. Opportunity, NASA's longest-running Mars rover, outlived its twin rover, Spirit, which went silent in 2010 after getting stuck in a sand trap and running out of power. Both rovers launched to the Red Planet in 2003 and are collectively known as NASA's Mars Exploration Rovers. The twin rovers made some remarkable scientific discoveries over the course of their missions on the Red Planet. Here we'll focus on some of the stranger things Opportunity and Spirit have spotted on Mars.

The twin Spirit and Opportunity rovers both landed on Mars in 2004

Credit: NASA/JPL-Caltech

1 JELLY DOUGHNUT

A mysterious object that bears a striking resemblance to a powdered, jelly-filled doughnut mysteriously appeared in front of the Opportunity rover's panoramic camera on January 8, 2014. Previous images of that exact same spot were suspiciously doughnut-free.

No one knew what it was or where it came from, but people on Earth were captivated by the Martian doughnut – you could say... they were eating it up!

After several weeks of analyzing the images taken by Opportunity, NASA scientists determined that what looked like a jelly-filled pastry was actually just a plain old rock that had been kicked up and displaced by the rover's wheels as it drove around.

2 A MARTIAN BUNNY RABBIT

NASA's Opportunity rover never found life on Mars, but in a photo of Meridiani Planum, it did find a mysterious object that looks like a long-eared bunny rabbit. The image was released in March 2004 (about two months after Opportunity arrived at the Red Planet) as part of the rover's "mission success" panorama. The bunny's ears appeared to move slightly in the weak Martian wind, so it couldn't have been a rock. NASA engineers said it appears to be "a piece of soft material that definitely came from our vehicle," like cotton insulation or a Vectran cover.

Credit: NASA

3 BLUEBERRIES ON MARS

Just a few months after the Opportunity rover arrived at the Red Planet, its cameras spotted this deceptively delicious-looking patch of rocks near the landing site. Not knowing exactly what they were looking at, scientists starting calling these strangely spherical rocks "blueberries." It's unclear exactly how these rocks came to be, but scientists believe that they constitute one of the earliest pieces of evidence that Mars had a very watery past.

Credit: NASA/JPL-Caltech/Cornell/USGS

Credit: NASA/JPL-Caltech/Cornell University

4 A PERSON ON MARS?

Opportunity's twin, a nearly identical rover named Spirit, also made some strange discoveries during its time roving around the Red Planet. In 2007, Spirit snapped a photo of what vaguely resembles a human-like figure perched on a rock. While some interpreted the image to be evidence of life on Mars, NASA has assured everyone that the "figure" was just a rock.

5 THE FIRST EXTRATERRESTRIAL METEORITE

On Jan. 6, 2005, the Opportunity rover found a basketball-size meteorite on Mars – the first meteorite ever discovered on another planet. Opportunity's spectrometers scoped out the space rock and determined that it's mostly made of iron and nickel. NASA named the meteorite Heat Shield Rock because it was spotted near Opportunity's heat shield, which was discarded during the rover's landing in 2003.

Credit: NASA/JPL/Cornell

JAMES WEBB SPACE TELESCOPE

ALL ABOUT NASA'S SUCCESSOR TO HUBBLE

WORDS: ELIZABETH HOWELL

NASA's James Webb Space Telescope, scheduled for launch in 2021, will probe the cosmos to uncover the history of the universe from the Big Bang to alien planet formation and beyond. It will focus on four main areas: first light in the universe, assembly of galaxies in the early universe, birth of stars and protoplanetary systems, and planets (including the origins of life.)

The James Webb Space Telescope (JWST) will launch on an Ariane 5 rocket from French Guiana, then take 30 days to fly a million miles to its permanent home: a Lagrange point, or a gravitationally stable location in space. It will orbit around L2, a spot in space near Earth that lies opposite from the sun. This has been a popular spot for several other space telescopes, including the Herschel Space Telescope and the Planck Space Observatory.

DID YOU KNOW...?

JWST will have a mission lifetime of at least 5 years, but it is expected to operate for 10+ years

The powerful $8.8 billion spacecraft is also expected to take amazing photos of celestial objects like its predecessor, the Hubble Space Telescope. Luckily for astronomers, the Hubble Space Telescope remains in good health and it's probable that the two telescopes will work together for JWST's first years. JWST will also look at exoplanets that the Kepler Space Telescope found, or follow up on real-time observations from ground space telescopes.

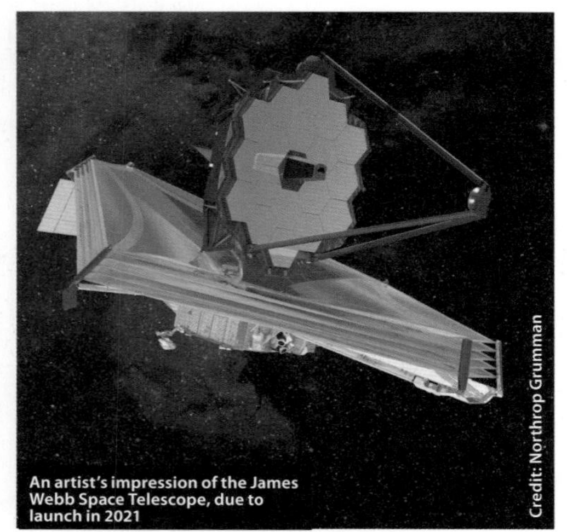

An artist's impression of the James Webb Space Telescope, due to launch in 2021

Credit: Northrop Grumman

JWST SCIENCE

JWST's science mandate is principally divided among four areas:

First light and reionization: This refers to the early stages of the universe after the Big Bang started the universe as we know it today. In the first stages after the Big Bang, the universe was a sea of particles (such as electrons, protons and neutrons), and light was not visible until the universe cooled enough for these particles to begin combining. Another thing JWST will study is what happened after the first stars formed; this era is called "the epoch of reionization" because it refers to when neutral hydrogen was reionized (made to have an electric charge again) by radiation from these first stars.

Assembly of galaxies: Looking at galaxies is a useful way to see how matter is organized on

NASA technicians moving the JWST's primary mirror in a clean room at the Goddard Space Flight Center in 2017

Credit: NASA/Desiree Stover

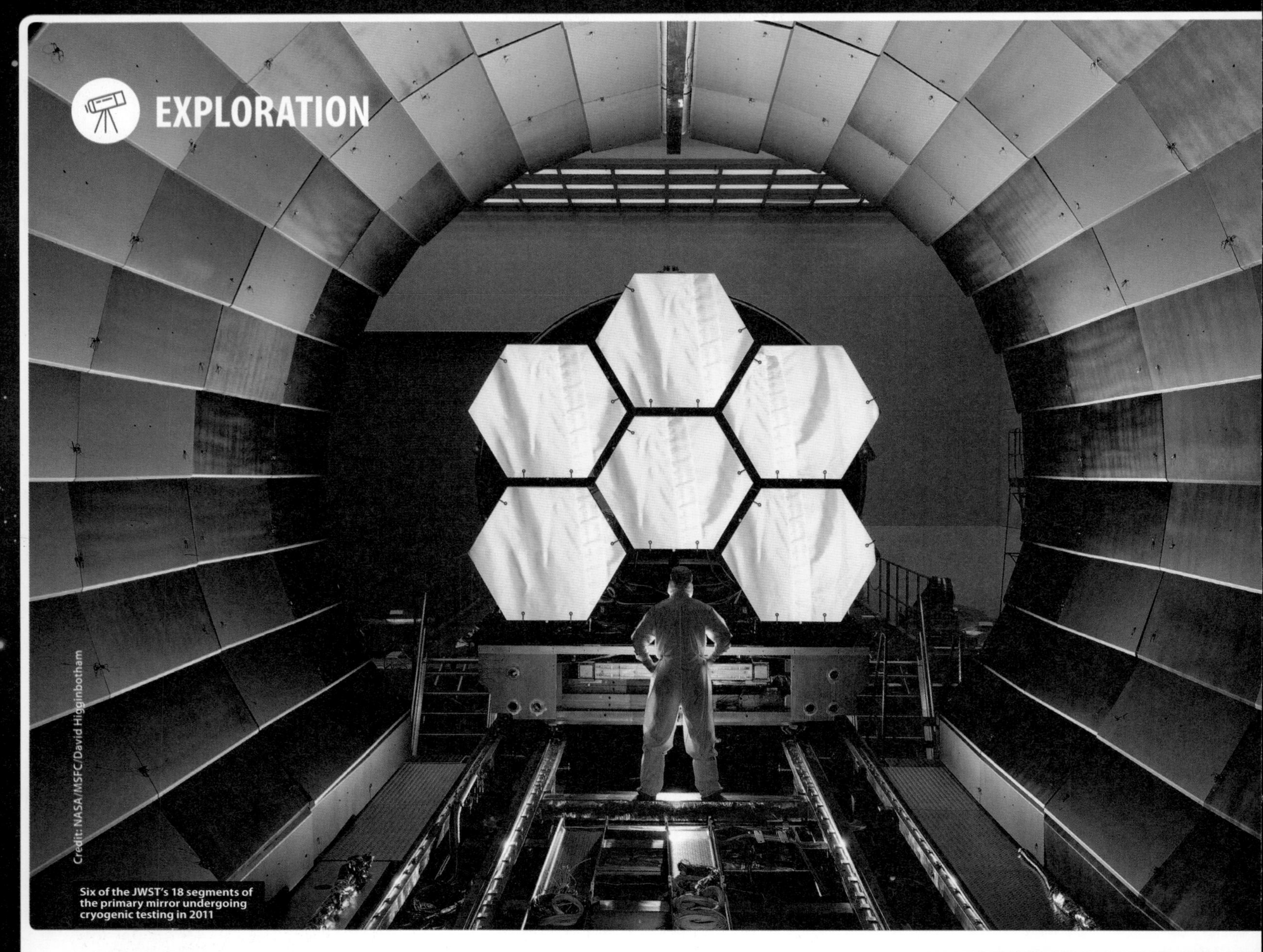
Credit: NASA/MSFC/David Higginbotham

Six of the JWST's 18 segments of the primary mirror undergoing cryogenic testing in 2011

"JWST'S INFRARED EYES WILL BE ABLE TO LOOK AT SOURCES OF HEAT, INCLUDING STARS THAT ARE BEING BORN"

gigantic scales, which in turn gives us hints as to how the universe evolved. The spiral and elliptical galaxies we see today actually evolved from different shapes over billions of years, and one of JWST's goals is to look back at the earliest galaxies to better understand that evolution. Scientists are also trying to figure out how we got the variety of galaxies that are visible today, and the current ways that galaxies form and assemble.

Birth of stars and protoplanetary systems: The Eagle Nebula's "Pillars of Creation" are some of the most famous birthplaces for stars. Stars come to be in clouds of gas, and as the stars grow, the radiation pressure they exert blows away the cocooning gas (which could be used again for other stars, if not too widely dispersed.) However, it's difficult to see inside the gas. JWST's infrared eyes will be able to look at sources of heat, including stars that are being born in these cocoons.

Planets and origins of life: The last decade has seen vast numbers of exoplanets discovered, including with NASA's planet-seeking Kepler Space Telescope. JWST's powerful sensors will be able to peer at these planets in more depth, including (in some cases) imaging their atmospheres. Understanding the atmospheres and the formation conditions for planets could help scientists better predict if certain planets are habitable or not.

INSTRUMENTS ON BOARD

The JWST will be equipped with four instruments:

- **Near-Infrared Camera (NIRCam)**
  Provided by the University of Arizona, this infrared camera will detect light from stars in nearby galaxies and stars within the Milky Way. It will also search for light from stars and galaxies that formed in the early universe. NIRCam will be outfitted with coronagraphs that can block a

Credit: NASA

The JWST is the most ambitious space telescope ever constructed bright object's light, making dim objects near those stars (like planets) visible.

- **Near-Infrared Spectrograph (NIRSpec)**
  NIRSpec will observe 100 objects simultaneously, searching for the first galaxies that formed after the Big Bang. NIRSpec was provided by the European Space Agency with help from NASA's Goddard Space Flight Center.
- **Mid-Infrared Instrument (MIRI)**
  MIRI will produce amazing space photos of distant celestial objects, following in Hubble's tradition of astrophotography. The spectrograph that is a part of the instrument will allow scientists to gather more physical details about distant objects in the universe. MIRI will detect distant galaxies, faint comets, forming stars and objects in the Kuiper Belt. MIRI was built by the European Consortium with the European Space Agency and NASA's Jet Propulsion Laboratory.
- **Fine Guidance Sensor/Near InfraRed Imager and Slitless Spectrograph (FGS/NIRISS)**
  This Canadian Space Agency-built instrument is more like two instruments in one. The FGS component is responsible for keeping the JWST pointed in exactly the right direction during its investigations. NIRISS will scan the cosmos to find signatures of the first light in the universe and seek out and characterize alien planets.

The telescope will also sport a tennis court-size sunshield and a 21.3-foot (6.5 meters) mirror – the largest mirror ever launched into space. Those components will not fit into the rocket launching the JWST, so both will unfurl once the telescope is in space.

James Webb's official NASA portrait, 1966

Credit: NASA

JAMES WEBB

The JWST is named after the former NASA chief James Webb. Webb took charge of the space agency from 1961 to 1968, retiring just a few months before NASA put the first man on the moon.

Although Webb's tenure as NASA administrator is most closely associated with the Apollo program, he is also considered a leader in space science. Even in a time of great political turmoil, Webb set NASA's science objectives, writing that the launch of a space telescope should be a key goal of the space agency.

NASA launched more than 75 space science missions under Webb's guidance. These included missions that studied the sun, stars and galaxies as well as space directly above Earth's atmosphere.

Kepler-186f was the first confirmed Earth-size exoplanet orbiting its star in the habitable zone

Credit: NASA/Ames/SETI Institute/JPL-Caltech

EXOPLANETS

WHAT DO WE KNOW ABOUT THE WORLDS BEYOND OUR OWN SOLAR SYSTEM?

WORDS: ELIZABETH HOWELL

Exoplanets are planets beyond our own solar system. Thousands have been discovered in the past two decades, mostly with NASA's Kepler Space Telescope

These worlds come in a huge variety of sizes and orbits. Some are gigantic planets hugging close to their parent stars; others are icy, some rocky. NASA and other agencies are looking for a special kind of planet: one that's the same size as Earth, orbiting a sun-like star in the habitable zone.

The habitable zone is the range of distances from a star where a planet's temperature allows liquid water oceans, critical for life on Earth. The earliest definition of the zone was based on simple thermal equilibrium, but current calculations of the habitable zone include many other factors, including the greenhouse effect of a planet's atmosphere. This makes the boundaries of a habitable zone "fuzzy." Astronomers announced in August 2016 that they might have found such a planet orbiting Proxima Centauri. The newfound world, known as Proxima b, is about 1.3 times more massive than Earth, which suggests that the exoplanet is a rocky world, researchers said. The planet is also in the star's habitable zone, just 4.7 million miles (7.5 million kilometers) from its host star. It completes one orbit every 11.2 Earth-days. As a result, it's likely that the exoplanet is tidally locked, meaning it always shows the same face to its host star, just as the moon shows only one face (the near side) to Earth.

Most exoplanets have been discovered by the Kepler Space Telescope, an observatory that began work in 2009 and was retired in October 2018, once it ran out of fuel. By the time of its retirement, Kepler had discovered over 2,600 confirmed exoplanets and revealed the existence of perhaps thousands of others. The total number of exoplanets discovered by all observatories (as of May 2019) is 4,065.

EARLY DISCOVERIES

While exoplanets were not confirmed until the 1990s, for years beforehand astronomers were convinced they were out there. That wasn't just wishful thinking, but rather because of how slowly our own sun and other stars like it spin, University of British Columbia astrophysicist Jaymie Matthews told Space.com. Matthews, who is the principal investigator of occasional exoplanet telescope observer MOST (Microvariability and Oscillations of STars), was involved in some of the earliest exoplanet discoveries.

An artist's impression of the surface of exoplanet Trappist-1f

Credit: NASA/JPL-Caltech

"BY THE TIME OF ITS RETIREMENT, KEPLER HAD DISCOVERED OVER 2,600 CONFIRMED EXOPLANETS"

Astronomers are searching for potentially habitable Earth-like worlds in other solar systems

Credit: NASA/JPL-Caltech/R. Hurt (SSC-Caltech)

Astronomers had an origin story for our solar system. Simply put, a spinning cloud of gas and dust (called the protosolar nebula) collapsed under its own gravity and formed the sun and planets. As the cloud collapsed, conservation of angular momentum meant the soon-to-be-sun should have spun faster and faster. But, while the sun contains 99.8 percent of the solar system's mass, the planets have 96 percent of the angular momentum. Astronomers asked themselves why the sun rotates so slowly.

The young sun would have had a very strong magnetic field, whose lines of force reached out into the disk of swirling gas from which the planets would form. These field lines connected with the charged particles in the gas, and acted like anchors, slowing down the spin of the forming sun and spinning up the gas that would eventually turn into the planets. Most stars like the sun rotate slowly, so astronomers inferred that the same "magnetic braking" occurred for them, meaning that planet formation must have occurred for them. The implication: Planets must be common around sun-like stars.

For this reason and others, astronomers at first restricted their search for exoplanets to stars similar to the sun, but the first two discoveries were around a pulsar (rapidly spinning corpse of a star that died as a supernova) called PSR 1257+12, in 1992. The first confirmed discovery of a world orbiting a sun-like star, in 1995, was 51 Pegasi b – a Jupiter-mass planet 20 times closer to its sun than we are to ours. That was a surprise. But another oddity popped up seven years earlier that hinted at the wealth of exoplanets to come.

A Canadian team discovered a Jupiter-size planet around Gamma Cephei in 1988, but because its orbit was much smaller than Jupiter's, the scientists did not claim a definitive planet detection. "We weren't expecting planets like that. It was different enough from a planet in our own solar system that they were cautious," Matthews said.

Most of the first exoplanet discoveries were huge Jupiter-size (or larger) gas giants orbiting close to their parent stars. That's because astronomers were relying on the radial velocity technique, which measures how much a star "wobbles" when a planet or planets orbit it. These large planets close in produce a correspondingly big effect on their parent star, causing an easier-to-detect wobble.

Before the era of exoplanet discoveries, instruments could only measure stellar motions down to a kilometer per second, too imprecise to detect a wobble due to a planet. Now, some instruments can measure velocities as low as a centimeter per second, according to Matthews. "Partly due to better instrumentation, but also because astronomers are now more experienced in teasing subtle signals out of the data."

DID YOU KNOW...?

Since the early 1990s, the number of known exoplanets has doubled approximately every 27 months

KEPLER, TESS AND OTHER OBSERVATORIES

Kepler launched in 2009 on a prime mission to observe a region in the Cygnus constellation. Kepler performed that mission for four years – double its initial mission lifetime – until most of its reaction wheels (pointing devices) failed. NASA then put Kepler on a new mission called K2, in which Kepler

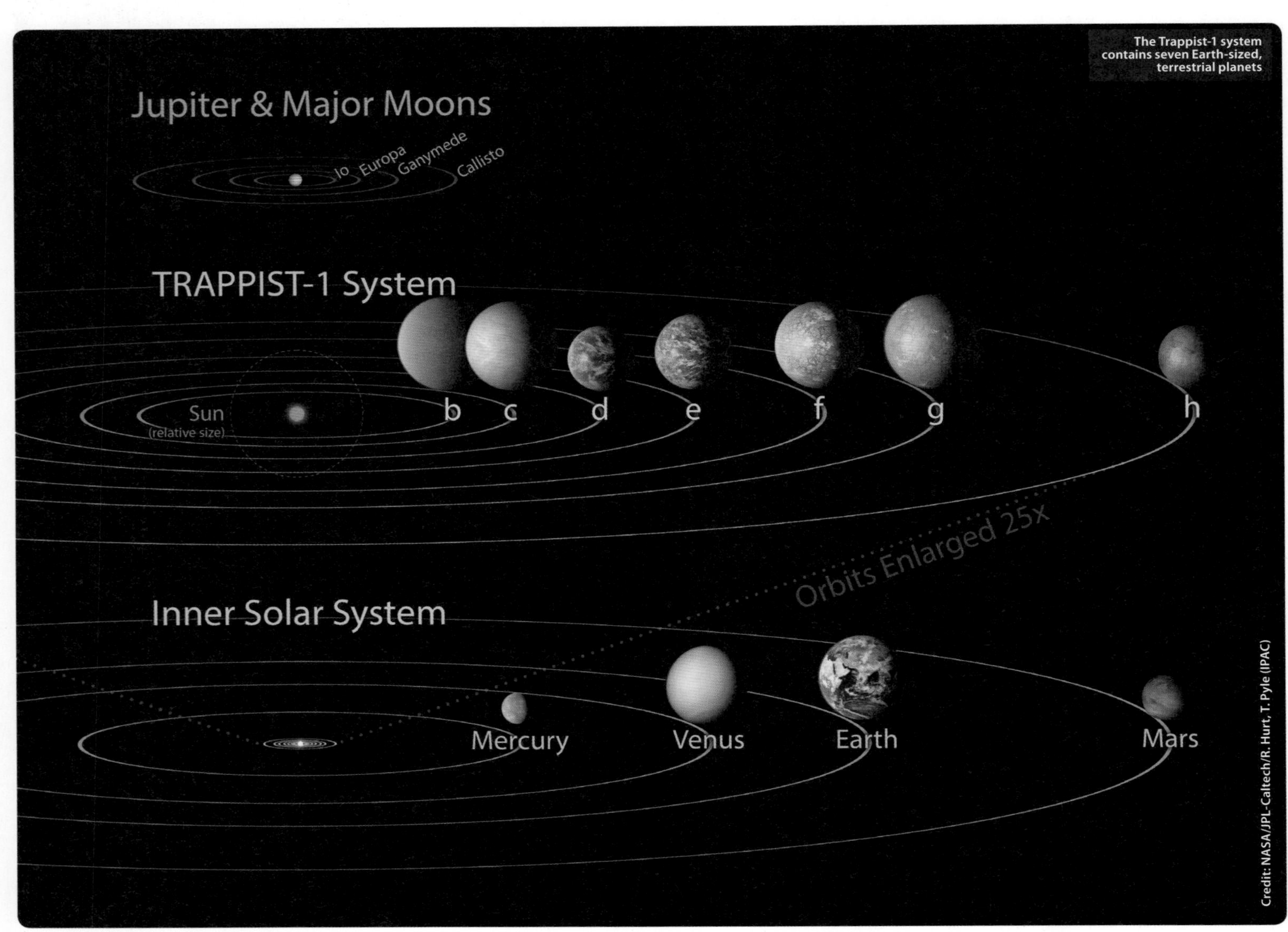

The Trappist-1 system contains seven Earth-sized, terrestrial planets

Credit: NASA/JPL-Caltech/R. Hurt, T. Pyle (IPAC)

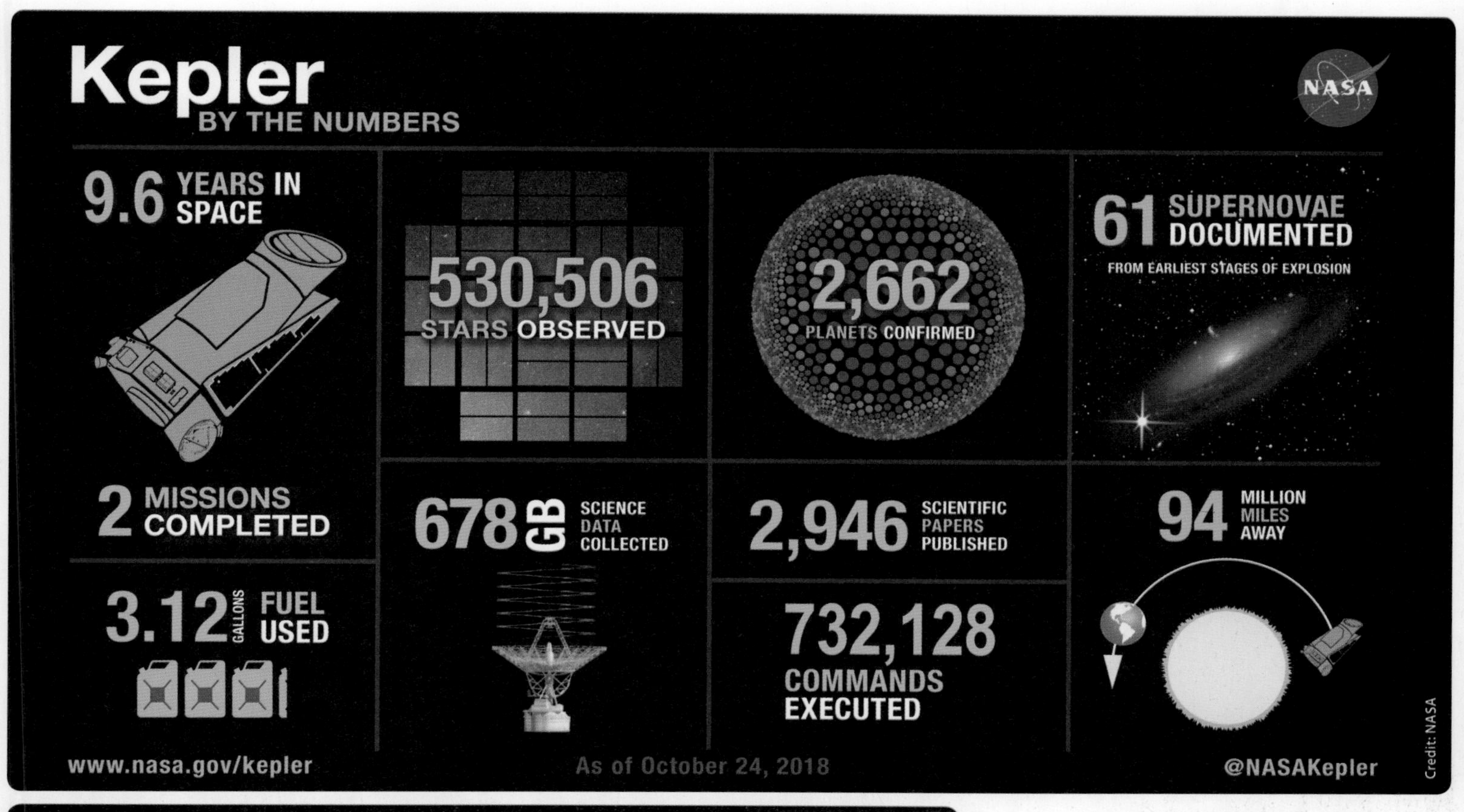

Credit: NASA

Hot Jupiters are among the most common exoplanets in the galaxy

Credit: ESA/Hubble & NASA

uses the pressure of the solar wind to maintain position in space. The observatory periodically switches its field of view to avoid the sun's glare. Kepler's pace of planetary discovery slowed after switching to K2, but it is still found hundreds of exoplanets using the new method. Its latest data release, in February 2018, contained 95 new planets.

Kepler has revealed a cornucopia of different types of planets. Besides gas giants and terrestrial planets, it has helped define a whole new class known as "super-Earths": planets that are between the size of Earth and Neptune. Some of these are in the habitable zones of their stars, but astrobiologists are going back to the drawing board to consider how life might develop on such worlds. Kepler's observations showed that super-Earths are abundant in our universe. Oddly, our solar system doesn't appear to contain a planet of that size, although it is theorised that a large planet nicknamed "Planet Nine" may be lurking beyond Neptune.

Kepler's primary method of searching for planets is the "transit" method. Kepler monitors a star's

"KEPLER HAS REVEALED A CORNUCOPIA OF DIFFERENT PLANETS"

light. If the light dims at regular and predictable intervals, that suggests a planet is passing across the face of the star. In 2014, Kepler astronomers (including Matthews' former student Jason Rowe) unveiled a new "verification by multiplicity" method that increased the rate at which astronomers promote candidate planets to confirmed planets. The technique is based on orbital stability – many transits of a star occurring with short periods can only be due to planets in small orbits, since multiply eclipsing stars that might mimic would gravitationally eject each other from the system in just a few million years.

While Kepler wrapped up its mission, a new observatory called the Transiting Exoplanet Survey Satellite (TESS) launched in April 2018. TESS will orbit the Earth every 13.7 days and will perform an all-sky survey over two years. It will survey the Southern Hemisphere in its first year, and the Northern Hemisphere (which includes the original Kepler field) in its second. The observatory is expected to reveal many more exoplanets, including at least 50 that are around the size of Earth.

NOTABLE EXOPLANETS

WITH THOUSANDS TO CHOOSE FROM, IT'S HARD TO NARROW DOWN A FEW. SMALL SOLID PLANETS IN THE HABITABLE ZONE ARE AUTOMATICALLY STANDOUTS, BUT MATTHEWS SINGLED OUT FOUR OTHER EXOPLANETS THAT HAVE EXPANDED OUR PERSPECTIVE ON HOW PLANETS FORM AND EVOLVE

51 PEGASI B

As mentioned earlier, this was the first planet to be confirmed around a sun-like star. Half the mass of Jupiter, it orbits around its sun at roughly the distance of Mercury from our Sun. 51 Pegasi b is so close to its parent star that it is likely tidally locked, meaning one side always faces the star.

Credit: ESO/M. Kornmesser/Nick Risinger (skysurvey.org)

PROMINENT PLANET-HUNTING OBSERVATORIES PAST AND PRESENT

- The HARPS spectograph on the European Southern Observatory's La Silla 3.6-meter telescope in Chile, whose first light was in 2003. The instrument is designed to look at the wobbles that a planet induces in a star's rotation. HARPS has found well over 100 exoplanets itself, and has regularly been used to confirm observations made by Kepler and other observatories.
- The Canadian Microvariability and Oscillations of STars (MOST) telescope, which started observations in 2003. MOST is designed to observe a star's astroseismology, or starquakes. But it also has participated in exoplanet discoveries, such as finding the exoplanet 55 Cancri e.
- The French Space Agency's CoRoT (COnvection ROtation and planetary Transits), operated between 2006-2012. It found a few dozen confirmed planets, including COROT-7b – the first exoplanet that were predominantly composed of rock or metal.
- The NASA/European Space Agency Hubble and NASA Spitzer space telescopes, which periodically observe planets in visible or infrared wavelengths, respectively. (More information about a planet's atmosphere is available in infrared.)

NASA's Spitzer telescope has been scanning the skies in infrared, providing some insight into the atmospheres of exoplanets

Credit: NASA/JPL-Caltech/R. Hurt (SSC)

- The European CHaracterising ExOPlanets Satellite (CHEOPS), which is expected to be ready for launch in late 2019. The mission is designed to calculate the diameters of planets accurately, particularly those planets that fall between super-Earth and Neptune masses.
- The NASA James Webb Space Telescope, which is expected to launch in 2021. It is specialized to observe in infrared wavelengths. The powerful observatory is expected to reveal more about the habitability of certain exoplanets' atmospheres.
- The European Space Agency's PLAnetary Transits and Oscillations of stars (PLATO) telescope, which is expected to launch in 2024. It is designed to learn how planets form and which conditions, if any, could be favorable for life..
- The ESA ARIEL (Atmospheric Remote-sensing Infrared Exoplanet Large-survey) mission, which is planned to launch in mid-2028. It is being designed to observe some 1,000 exoplanets and will also perform a survey of the chemical compositions of their atmospheres.

Credit: NASA

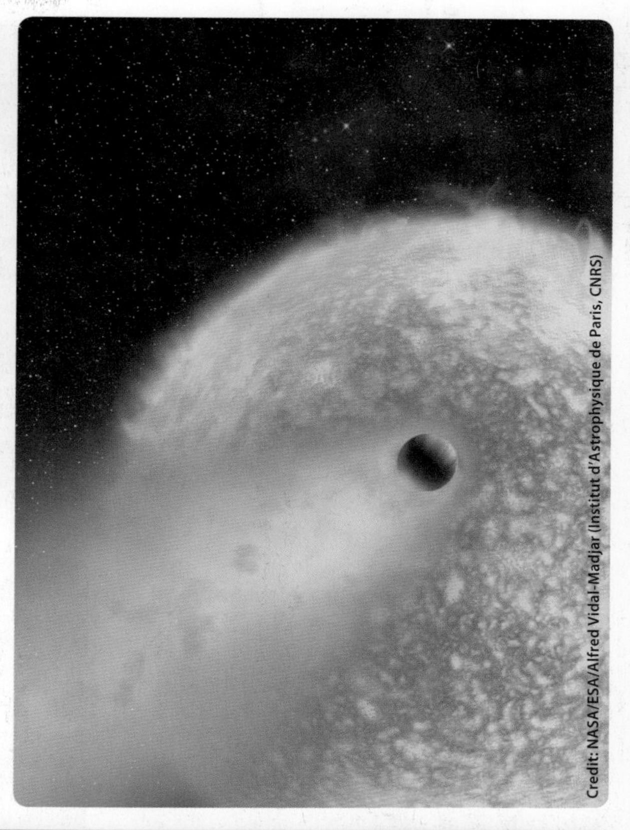

Credit: NASA/ESA/Alfred Vidal-Madjar (Institut d'Astrophysique de Paris, CNRS)

ABOVE:

WASP-33B

This planet was discovered in 2011 and has a sort of "sunscreen" layer – a stratosphere – that absorbs some of the visible and ultraviolet light from its parent star. Not only does this planet orbit its star "backward," but it also triggers vibrations in the star, seen by the MOST satellite.

RIGHT:

HD 209458 B

This was the first planet found (in 1999) to transit its star (although it was discovered by the Doppler wobble technique) and in subsequent years more discoveries piled up. It was the first planet outside the solar system for which we could determine aspects of its atmosphere, including temperature profile and the lack of clouds. Matthews participated in some of the observations using MOST.

Credit: NASA/JPL-Caltech

55 CANCRI E

This super-Earth orbits a star that is bright enough to see by eye, meaning astronomers can study the system in more detail than almost any other. Its "year" is only 17 hours and 41 minutes long (recognized when MOST gazed at the system for two weeks in 2011). Theorists speculate that the planet may be carbon-rich, with a diamond core.

THE NEW SPACE RACE

HOW WILL PRIVATE SPACE TRAVEL TRANSFORM NASA'S NEXT 60 YEARS?

WORDS: DORIS ELIN SALAZAR

NASA's next 60 years will probably be very different than its first six decades. When the agency opened for business in 1958, private spaceflight was just a sci-fi dream. But companies such as Elon Musk's SpaceX and Jeff Bezos' Blue Origin are working to make that dream a reality and open the space frontier to huge numbers of people for the first time. What role will NASA play in the private sector's liftoff? Space.com recently talked to three commercial-spaceflight experts to get some ideas.

First, people should understand that about 75 percent of the worldwide space enterprise is already commercial, said Scott Hubbard, an adjunct professor in the Department of Aeronautics and Astronautics at Stanford University. This includes the satellites belonging to DirecTV and Sirius XM radio. What's new s"is the extension of that into the human realm," said Hubbard, who also previously directed NASA's Ames Research Center in Silicon Valley. He served as the agency's "Mars czar," restructuring NASA's robotic Red Planet-exploration program after it suffered several failures in the 1990s. And if private companies can get the price of a suborbital flight down to about $50,000, "you get a lot of interest," Hubbard told Space.com.

The highest-profile program currently in the works between NASA and the private sector is the agency's Commercial Crew Program, said Eric Stallmer, president of the nonprofit Commercial Spaceflight Federation.

Commercial Crew is encouraging the development of U.S. spacecraft that will carry astronauts to and from the International Space Station (ISS). Toward this end, NASA has awarded multibillion-dollar contracts to both SpaceX and Boeing, which are building capsules called Crew Dragon and CST-100 Starliner, respectively. These craft are currently scheduled to start flying astronauts in 2019.

There's also the maturing commercial cargo program, which has given contracts to SpaceX and Northrop Grumman Corp. to fly robotic cargo missions to the ISS. Both of these companies have already completed numerous such flights.

"NASA'S BUDGET IS ABOUT FIVE TIMES LARGER THAN THE NEXT BIGGEST SPACE AGENCY"

Both Hubbard and Stallmer believe that NASA wins by relying on private industry to provide such services in low Earth orbit. Hubbard argued that this strategy allows the space agency to continue "exploring the fringe where there really is no business case."

NASA has a budget about five times larger than the next biggest national space agency out there, but the U.S. agency's ambitious goals are still costly, said Stallmer. To get the most bang for the buck, "you'd have to leverage the innovation and technology that is in the private sector and let NASA do the exquisite" projects. The "exquisite" projects, Stallmer explained, are the "push-the-envelope-type things on deeper space exploration."

"I see it not only as a cooperation or a collaboration, but maybe even interdependence," Hubbard said. "Without a thriving spaceflight entrepreneurship sector, I don't think that deep-space exploration with [regular] people is sustainable," he added. "And I think using the way in which the private sector has demonstrated they can reduce costs, through more nearly assembly-line production techniques, is really critical to sustainable space exploration in the future."

Phil McAlister, director of commercial spaceflight at NASA, also advocated these public-private partnerships. Private companies offer the advantages

Northrop Grumman's Cygnus cargo spacecraft has made several resupply missions to the ISS

DID YOU KNOW...?

A 2017 study found that SpaceX cargo launches save NASA around $183,000 per kilogram

In 2018 NASA announced their first astronauts who will fly on the upcoming SpaceX and Boeing commercial spacecraft missions to the ISS

NASA's future Mars ambitions may come to rely on continued cooperation with private companies

uick, being nimble, being fast, making
naybe without perfect knowledge –
g forward and adjusting as required,"
old Space.com.
icials, McAlister said, "have a lot of
. a lot of discussions, and things tend to
" than in private industry. "The private
ting to move fast and wanting to be
ve and NASA having our 50 years of
ceflight experience… you bring those
together, and they actually complement
very effectively," he explained.
e are more players within that private
"pie" now than there used to be,
d. Aerospace giants like Lockheed
ing and Northrop Grumman Corp. build
or NASA and the U.S. National Oceanic
pheric Administration (NOAA), but they'll
ontinue pursuing big-dollar defense
hese standard government contractors
ger the only options that NASA can
n.
the future," Stallmer said, "the contracts
cally went to the big three or the big
ing elsewhere. And you're seeing smaller,
le companies entering the marketplace
eting for a lot of this work. So, it won't
r standard government contractors… it's
ger pool… to choose from."
r also said there's now a big shift in
and operates spacecraft, as a result of
ment spaceflight customers. "I think the
of nongovernment customers only
red in the last 10 or 15 years in the
stry," he said. "Prior to that, it was pretty
NASA and governments [that] were
ners, and when you have that kind of
[it] makes sense for [NASA] to own and
e hardware."
Alister added, "when you have the
y for other customers, then it makes
hift some of that responsibility of
ent to the companies, to the private
w them to own and operate their
and then they can sell it to other
and that brings the cost down for NASA
erybody, because they can advertise their
over a larger customer base." He called
of a win-win scenario."
l these future spaceflight customers
h, at least in the near term. After all,
ace travel, even to the nearby suborbital
likely remain quite expensive for a while,
ve said.
doesn't mean the rest of us have no
y in the ongoing private-spaceflight
"I think we're going to need a lot of
eople," Stallmer mused. "We're going
lot of builders… not just aerospace
anymore. It's artisans, people that can
ands."

In 2016, NASA selected six private companies to help develop deep space habitat concepts and prototypes

As well as technological innovation, stunts like SpaceX's "Starman" launch have also helped reinvigorated public interest in space exploration

The development of reusable rockets greatly reduces the cost of launches

SPACE CALENDAR 2021

LAUNCHES, SKY EVENTS AND MORE – WHAT TO LOOK FORWARD TO IN THE COMING MONTHS

WORDS: HANNEKE WEITERING

*Additional scheduled events sourced from Spaceflight Now

APRIL

Credit: X-ray: NASA/Joel Kowsky

APRIL 7

A SpaceX Falcon 9 rocket will launch about 60 satellites for SpaceX's Starlink broadband network. It will lift off from Space Launch Complex 40 at Cape Canaveral Space Force Station in Florida at 12:34 p.m. EDT (4:34 p.m. GMT).

Credit: Roscosmos/Oleg Novitsky/Luc van den Abeelen

APRIL 9

A Russian Soyuz rocket will launch the crewed Soyuz MS-18 spacecraft to the International Space Station with NASA astronaut Mark Vande Hei and two Russian cosmonauts, Oleg Novitsky and Pyotr Dubrov.

APRIL 11 NEW MOON

Credit: NASA/GCTC/Andrey Shelepin

APRIL 16-17

NASA astronaut Kate Rubins and Russian cosmonauts Sergey Ryzhikov and Sergey Kud-Sverchkov will return to Earth from the International Space Station in their Soyuz MS-17 spacecraft.

Credit: China News Service, CC BY 3.0

APRIL 29

A Chinese Long March 5B rocket will launch Tianhe 1, the core module for a Chinese space station low Earth orbit. It will lift off from the Wenchang Space Launch Center in China's Hainan province.

Also scheduled to launch in April*:

- The SpaceX Crew-1 mission will return to Earth with NASA astronauts Michael Hopkins, Victor Glover and Shannon Walker, and JAXA astronaut Soichi Noguchi, in late April or early May

MAY

MAY 4 STAR WARS DAY (MAY THE FOURTH BE WITH YOU)

MAY 4-5

The Eta Aquarid meteor shower, which is active from mid-April all the way to the end of May, peaks overnight.

MAY 15

Mercury reaches its highest point in the evening sky, shining brightly at magnitude 0.3. See it just above the western horizon right after sunset.

MAY 17

A United Launch Alliance Atlas V rocket will launch the U.S. Space Force's fifth Space Based Infrared System Geosynchronous satellite (SBIRS GEO 5) from Space Launch Complex 41 at Cape Canaveral Space Force Station in Florida.

Credit: NASA

MAY 26

The full moon of May, known as the Full Flower Moon, arrives at 7.14am EDT (11.14am GMT). It will also be the closest "supermoon" of the year. That night, a total lunar eclipse, also known as a "Blood Moon," will be visible from Australia, parts of the western United States, western South America and Southeast Asia.

Credit: Giuseppe Donatiello, CC0

Also scheduled to launch in May*:

- Starliner OFT-2: A United Launch Alliance Atlas V rocket will launch Boeing's CST-100 Starliner spacecraft on its second uncrewed mission to the International Space Station, following a partial failure in December 2019. The Orbital Flight Test 2 (OFT-2) mission will lift off from Space Launch Complex 41 at Cape Canaveral Space Force Station in Florida.
- Arianespace will use an Ariane 5 ECA rocket, designated VA254, to launch the Star One D2 and Eutelsat Quantum communications satellites from the Guiana Spaceport near Kourou, French Guiana.
- Arianespace will use a Soyuz rocket to launch 36 satellites into orbit for the OneWeb internet constellation. The mission, called OneWeb 7, will lift off from the Vostochny Cosmodrome in Siberia.
- China's Tianwen-1 Mars rover will touch down on the Red Planet.

JUNE

JUNE 1

A SpaceX Falcon 9 rocket will launch the Transporter 2 rideshare mission with several small satellites for commercial and government customers. It will lift off from Space Launch Complex 40 at Cape Canaveral Space Force Station in Florida.

Credit: NASA Kennedy from United StatesNASA/Tony Gray and Sandra Joseph

JUNE 10

An annular solar eclipse, also known as a "ring of fire" eclipse, will be visible from parts of Russia, Greenland and and northern Canada. Skywatchers in Northern Asia, Europe and the United States will see a partial eclipse.

Credit: Thinkstock

Credit: Getty Images

JUNE 24

The full moon of June, known as the Full Strawberry Moon, arrives at 2:40 p.m. EDT (1940 GMT).

JUNE 20

The solstice arrives at 11:16 p.m. EDT (0316 June 21 GMT), marking the first day of summer in the Northern Hemisphere and the first day of winter in the Southern Hemisphere.

Credit: Jessie Eastland, CC BY-SA 4.0

Also scheduled to launch in June*:

- A SpaceX Falcon 9 rocket will launch a Dragon cargo resupply mission (CRS-22) to the International Space Station. It will lift off from Launch Complex 39A at NASA's Kennedy Space Center in Florida.
- A U.S. Air Force and Northrop Grumman Minotaur 1 rocket will launch a classified spy satellite for the U.S. National Reconnaissance Office in a mission called NROL-111. It will lift off from Pad 0B at NASA's Wallops Flight Facility in Wallops Island, Virginia.
- A SpaceX Falcon 9 rocket will launch the Turksat 5B communications satellite from Cape Canaveral, Florida.
- A United Launch Alliance Atlas V rocket will launch the STP-3 rideshare mission for the U.S. Space Force. It will lift off from Cape Canaveral Space Force Station in Florida.

JULY

JULY 5 HAPPY APHELION DAY!
EARTH IS FARTHEST FROM THE SUN TODAY.

JULY 9
NEW MOON

JULY 12
Conjunction of the moon and Venus. The waxing crescent moon will pass about 3 degrees to the north of Venus

Credit: NASA/Johns Hopkins University Applied Physics Laboratory/Carnegie Institution of Washington

Credit: Gregory H. Revera, CC BY-SA 3.0

JULY 23
The full moon of July, known as the Full Buck Moon, arrives at 10:37 p.m. EDT (0237 July 24 GMT).

Credit: NASA / JPL / Space Science Institute

JULY 24
Conjunction of the moon and Saturn. The full moon will swing about 4 degrees to the south of Saturn in the dawn sky.

Also scheduled to launch in July*:

- A SpaceX Falcon Heavy rocket will launch the USSF-44 mission for the U.S. Air Force. The mission will lift off from NASA's Kennedy Space Center in Florida and is expected to deploy two undisclosed payloads into geosynchronous orbit.
- A SpaceX Falcon 9 rocket will the U.S. Space Force's fifth third-generation navigation satellite for the Global Positioning System (GPS 3 SV05). It will lift off from Space Launch Complex 40 at Cape Canaveral Space Force Station in Florida.
- India's Polar Satellite Launch Vehicle (PSLV) will launch the Indian RISAT 1A radar Earth observation satellite from the Satish Dhawan Space Center in Sriharikota, India.
- India's Small Satellite Launch Vehicle (SSLV) will launch its first commercial mission with four Earth observation satellites for the Seattle-based company BlackSky Global. It will lift off from the Satish Dhawan Space Center in Sriharikota, India.

AUGUST

AUG 2
Saturn at opposition. The ringed planet will be directly opposite the sun in Earth's sky around the same time that it makes its closest approach to Earth all year. This means it will appear at its biggest and brightest of the year. Saturn will reach its highest point in the night sky around midnight.

AUG 11-12
The annual Perseid meteor shower, which is active from mid-July to the end of August, peaks overnight.

Credit: Ahmed abd elkader mohamed, CC BY-SA 4.0

AUG 18
A SpaceX Falcon 9 rocket will launch a Dragon cargo resupply mission (CRS-23) to the International Space Station. It will lift off from Launch Complex 39A at NASA's Kennedy Space Center in Florida.

AUG 21
A Russian Soyuz rocket will launch the Progress 79 cargo resupply spacecraft to the International Space Station. It will lift off from the Baikonur Cosmodrome in Kazakhstan.

Credit: NASA/Crew of STS-132

Credit: Miyuki Meinaka, CC BY-SA 4.0

AUG 22
The full moon of August, known as the Full Sturgeon Moon, occurs at 8:02 a.m. EDT (1202 GMT). This will also be a so-called "Blue Moon" because it is the third full moon in a season that has four full moons.

Also scheduled to launch in August*:

- A United Launch Alliance Atlas V rocket will launch the USSF-8 mission for the Space Force's Geosynchronous Space Situational Awareness Program (GSSAP). It will lift off from Space Launch Complex 41 at Cape Canaveral Space Force Station in Florida.

SEPTEMBER

SEPT 6 NEW MOON

SEPT 14
Neptune at opposition. The gas giant will appear at its biggest and brightest of the year, shining at magnitude 7.8. (You'll need a telescope to see it.)

SEPT 13
A SpaceX Falcon 9 rocket will launch a Crew Dragon spacecraft on the Crew-3 mission, the third operational astronaut flight to the International Space Station. On board will be NASA astronauts Raja Chari and Thomas Marshburn, and European Space Agency astronaut Matthias Maurer. (The fourth crewmember has not yet been announced). It will lift off from Launch Complex 39A at NASA's Kennedy Space Center in Florida.

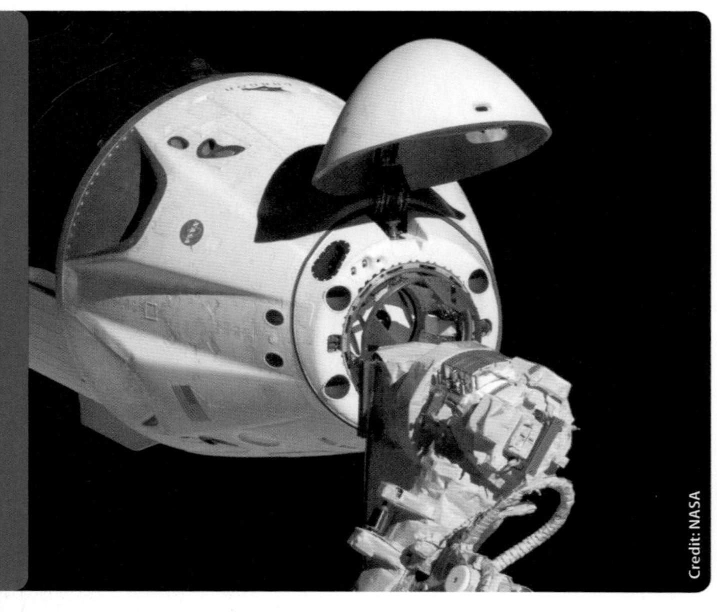

Credit: NASA

SEPT 20
The full moon of September, known as the Full Harvest Moon, occurs at 7:55 p.m. EDT (2355 GMT).

Credit: Thinkstock

Credit: Thinkstock

SEPT 22
The equinox arrives at 3:21 p.m. EDT (1921 GMT), marking the first day of autumn in the Northern Hemisphere and the first day of spring in the Southern Hemisphere.

Also scheduled to launch in September*:
- An Arianespace Soyuz rocket will launch two satellites for Europe's Galileo navigation constellation. It will lift off from the Guiana Space Center near Kourou, French Guiana.
- Boeing plans to launch the first crewed test flight of its Starliner spacecraft, which will send NASA astronauts Mike Fincke, Nicole Mann, and Barry "Butch" Wilmore to the International Space Station on an Atlas V rocket. The mission will lift off from Cape Canaveral Space Force Station in Florida.
- A SpaceX Falcon 9 rocket will launch the first two WorldView Legion Earth observation satellites from Vandenberg Air Force Base in California

OCTOBER

OCT 8

The Draconid meteor shower, which is active Oct. 6-10, will peak overnight.

OCT 16

NASA will launch its Lucy mission to study the Trojan asteroids. It will lift off from Kennedy Space Center in Florida on a United Launch Alliance Atlas V rocket.

Credit: NASA/Daniel Casper

OCT 21-22

The annual Orionid meteor shower, which is active all month long, peaks overnight.

Credit: Thinkstock

Credit: Getty Images

OCT 20

The full moon of October, known at the Full Hunter's Moon, occurs at 10:57 a.m. EDT (1457 GMT).

Credit: NASA

OCT 31

NASA's James Webb Space Telescope is scheduled to lift off from the Guiana Space Center in Kourou, French Guiana, on an Ariane 5 ECA rocket.

Also scheduled to launch in October*:

- A SpaceX Falcon Heavy rocket will launch the USSF 52 mission for the U.S. Space Force. It will lift off from Launch Complex 39A at NASA's Kennedy Space Center in Florida.
- The Soyuz MS-18 crew capsule will return to Earth from the International Space Station with Russian cosmonaut Oleg Novitsky, as well as two space tourists: Russian film director Klim Shipenko and a (not-yet-named) Russian actress, who will have arrived on the Soyuz MS-19 mission in September and plan to film a movie in space.